BEST RESUMES FOR SCIENTISTS AND ENGINEERS

SECOND EDITION

Adele Lewis
David J. Moore

John Wiley & Sons, Inc.

New York · Chichester · Brisbane · Toronto · Singapore

Figures 13.3–13.7 are used by permission of Symantec Corporation.

This text is printed on acid-free paper.

Copyright © 1993 by John Wiley & Sons, Inc.

Library of Congress Cataloging-in-Publication Data

Lewis, Adele Beatrice, 1927–
    Best resumes for scientists and engineers / by Adele Lewis and
David J. Moore. (Second Edition)
        p.   cm.
    Includes index.
    ISBN 0-471-59452-0 (alk. paper). — ISBN 0-471-59451-2 (pbk. :
alk. paper)
    1. Scientists—Employment.   2. Engineers—Employment. 3. Résumés
(Employment)   I. Moore, David J. (David Jewel), 1936–
II. Title.
Q148.L47   1993
650.14'0245—dc20   93-2849

Printed in the United States of America

10 9 8 7 6 5 4 3 2 1

Preface

Scientists, engineers, computer specialists, and other technical professionals have historically found themselves in demand. Because they possess the expertise needed to develop and operate technology in the United States and worldwide, they have, for the most part, found themselves in the ranks of the employed. Before the 1990s, technical professionals were seldom without work. If they left a job, they easily found another. They were truly in a "buyer's market." Many scientists and engineers boasted that they never used a resume. Employers were begging for qualified people to fill jobs. They either recruited technical professionals directly into positions or, because of the scarcity of qualified candidates, overlooked the need for a resume.

Times have changed. The recession of the early 1990s and the dynamics of world politics have put millions of highly qualified, trained engineers, scientists, and other technical professionals in the ranks of the unemployed. For the first time, many have encountered the shock of not being able to step easily into another position. They have found themselves competing with others in similar circumstances—and these others are often equally or better qualified for the few positions available.

In reality, engineers and scientists have never been excused from developing resumes to assist them in job changes and job searches. Some may have found that prospective employers were not in a position to be choosy. However, while this may have made the transition from job to job appear easier, it did not necessarily enhance the career of the person changing jobs. If a high-quality resume had been used to gain access to a number of opportunities, the job seeker would have been better off because of the increased number of choices available. Without resumes, engineers or scientists looking for a change were limited to the employers who sought them out because of a need that likely bordered on desperation. While they were never without work, their careers suffered when measured against the choices and opportunities they never evaluated.

The economy and the demand for technical professionals may turn again in favor of job seekers. However, the lessons of massive layoffs, few opportunities, and keen competition should be written in the memory of each person who has experienced these traumas. Technical professionals should never again be without resumes designed to present their strengths in powerful and effective terms. They must be prepared!

The second edition of *Best Resumes for Scientists and Engineers* shows you, as a qualified engineer, scientist, or technical professional, how to prepare the most powerful and effective resume possible. This is done in a step-by-step manner using worksheets to help you organize your information. The worksheet is a tool that assists you in getting past the most difficult step in writing your resume— the first draft. This is the foundation upon which you will build a marketing document that will help you "sell" your background and abilities to a prospective employer and win a face-to-face interview. Given this opportunity, you can then secure the position.

To assist you in preparing your resume, sample resumes covering representative disciplines found in high-technology industries are included as models. It was impossible to include every job category for scientists, engineers, or technical professionals, but the book does provide a wide variety of examples generously salted with advice that comes from experience. The goal is *not* to show you your own resume, but to give you a tool that will enable you to craft the most effective resume possible.

The purpose of your resume is to represent you, your background, and your qualifications. Because it is the only representation of you that a prospective employer sees before making a decision, it must create an impact that shows you in the strongest, most positive way. Your resume must demonstrate instantly that you are a viable candidate with substantial qualifications. It must create an immediate desire to bring you in for a face-to-face interview.

Your resume is not likely to be the only one the prospective employer receives. If you have sent it in response to an advertisement or other solicitation, you may be in competition with many others— perhaps hundreds. Most employers spend from 10 to 20 seconds on each resume. Obviously, they do not read the entire resume in such a short time. What they do is quickly scan it looking for something that will catch their eye and cause them to schedule an interview.

Perhaps they are seeking technical "buzzwords," work experience, or education. Whatever it is, you should ensure that your resume says the things that will attract the reader.

Most experienced readers and preparers of resumes agree that the ideal length of a resume is one, or at the most two, pages. It is possible to have a longer resume, but the first page be must so powerful that it virtually "grabs" the reader and demands that the rest of the resume be read. Remember: You have 20 seconds at the most, so make your words count!

Typographical, grammatical, or spelling errors are sure to turn off resume reviewers. These are considered sloppy work, and they justify discarding the resume without further consideration. Other turnoffs include the use of poor-quality paper in weird colors, inclusion of photographs, and unclear and rambling writing.

Best Resumes for Scientists and Engineers, Second Edition, includes new concepts of job search that reflect the dynamics of the economy. It is, in fact, a completely new book. The goal remains the same: to assist technical professionals in the preparation and use of their resumes. However, it goes much further, providing insights and examples that will lead you step by step to the development of a resume that presents you in the most effective manner possible.

Resumes are included in this edition that span the spectrum of the technical environment, from people just entering the market or changing careers to top-level managers. The book also introduces the use of technologies that either were not available or were in limited use when the first edition was published. This includes the use of sophisticated word processors that make extensive use of automated spelling checkers, thesauruses, and grammar checkers. Word processing makes it possible to use a variety of type styles to enhance the appearance and attractiveness of the resume. Microsoft Corporation's *Word for Windows 2.0* was the word processor used for many of the resumes presented in the book, while Microsoft's *TrueType* and MicroLogic Corporation's *MoreFonts* were used to create various type styles and special effects.

The result is a guide book that will enable you to produce a resume that communicates that you know where you are going with your career and your background qualifies you for the position you seek.

This new, updated edition is based on the solid principles that made the first edition a classic, while adding the ingredients that make it the most up-to-date and effective book in its field.

DAVID J. MOORE

San Juan Capistrano, California
August 1993

Trademarks

Windows 3.1, *Word for Windows 2.0*, *Access 1.0*, and *Foxbase* are registered trademarks of Microsoft Corporation.

WordPerfect 5.2 and *WordPerfect for Windows* are registered trademarks of WordPerfect Corporation.

WordStar 7.0 is a registered trademark of WordStar International Incorporated.

Ami Pro is a registered trademark of Lotus Development Corporation.

PFS: Write is a trademark of Spinnaker Software.

Act! for Windows is a registered trademark of Symantec Corporation.

Paradox 4.0, *Paradox for Windows*, and *dBASE IV* are registered trademarks of Borland International.

PackRat is a trademark of Polaris Software Inc.

Grandview is a trademark of Symantec Corporation.

ResumeMaker with Career Planning is a trademark of Individual Software, Inc.

Career Design is a trademark of Career Design Software.

Office Accelerator is a trademark of Baseline Data Systems Corporation.

MoreFonts is a registered trademark of MicroLogic Software.

IBM is a registered trademark of International Business Machines Corporation.

Macintosh is a registered trademark of Apple Computer, Inc.

All other company and product names are trademarks or registered trademarks of their respective owners

Contents

Contents

1

Content and Style

THE RESUME DEFINED

A resume is defined by the *Random House Dictionary of the English Language* as "a summing up or summary"; or "a brief account of personal, educational, and professional qualifications and experience, as of an applicant for a job."

Actually, it is this and much more. The resume has come to be understood as the primary job search tool. When individuals decide to look for work or seek new jobs, the first positive action they take is to create or update their resumes. When employers begin to search for new employees, the first demand made of applicants is that they submit a resume.

The truth is, however, that the resume is overrated as a job search tool. It often does not reflect the abilities, character, or precise experience of the person it represents. Employers demand resumes of job candidates as if they were holy writs, but no employer would place sufficient faith in one to extend an offer to hire without first meeting the candidate face to face. In fact, although resumes are necessary tools in a job search, their value is questionable, unless they are structured to maximize their effectiveness.

To get the most from a resume you must understand what a resume really is, what it is supposed to do, and how to create a resume that achieves your job search objectives.

THE RESUME'S PURPOSE

The purpose of the resume is to present you favorably to a prospective employer so that employer will want to meet you in a face-to-face interview. Thus, a resume is a marketing tool that presents you and your background effectively. It is not a biography, nor must it contain every detail of your personal or work life. It must be factual and truthful, but you are under no obligation to provide information in it that may be harmful or detrimental to your chances for employment.

CONTENTS OF THE RESUME

Each resume must favorably describe the person it represents. Note the items listed below that a resume *must* include, *may* include, and *should not* include.

Must Include

- Name, address, and telephone number(s)
- Description of your work history
- Education, including degrees
- Awards, honors, or citations, both work-related and academic
- Publications and professional presentations

May Include

- Job objective or career goal statement
- Career summary
- Security clearance
- Memberships in professional organizations and offices held
- Foreign language competencies
- Military service
- Availability or willingness to travel or relocate

Should Not Include

- Salary information, including salary history, needs, or expectations
- Reasons for leaving past jobs or current position
- Personal information, including physical description, birthdate, marital status, names/ages of children or spouse, or religion
- Names or addresses of references or former employers
- Personal photographs

IDENTITY INFORMATION

Begin your resume with your name, address, and telephone numbers. Do not use any artwork or embellishments that might detract from a professional look. *Never* include a personal photograph. A photograph can be used to discriminate against you and open employers up to charges associated with such discrimination. Some employers are so sensitive about the issue that they will not consider any resume with a photograph attached. (See additional discussion in Chapter 7, Information Never to Include.)

Name

Use your formal name. A good rule is to use the same form as appears on your personal checks. You may use your full name complete with middle name or your middle initial. You may also choose not to use an initial. Do not use nicknames on your resume. You want to come across as businesslike as possible. For example, if your name is John Alan Williams, listing it as Jack Williams or John (Jack) Williams is too informal. If you changed your name during the history recorded on your resume, you may wish to note both names. However, because you will not be present to explain your name change when your resume is read, you may wish to include only your current name. Remember, however, that a prospective employer may contact someone at a previous employer who may not recognize your current name. If this is a concern, provide an explanation in your cover letter.

Address

Use a street address if possible. A post office box projects an image of impermanence or instability. Although this may not be the case for you, you will not be there to influence the reader.

Telephone Numbers

In most cases, use your home telephone number only. If you have a Fax at home, list that number also. Do not list a work number unless your current employer is aware of your job search.

RESUME STYLES

After deciding what information to include in your resume, you must choose a resume format or style. There are three basic resume styles: the historical/chronological, the functional, and the synoptic/amplified.

The style you choose does not alter the content of your resume; it merely rearranges the content in order to emphasize it. Although the pros, cons, and strengths of each format will be discussed here, you should note that 10 corporate executives surveyed unanimously chose the historical/chronological format as the preferred resume style. This is not meant to bias you in favor of this approach; it is a caveat: Carefully examine your reasons for using an alternate format.

Historical/Chronological Format

Resumes using the historical/chronological style or format present information chronologically, that is, in a historical style as the events being described occurred over time. However, the information is arranged in *inverse* chronological order beginning with the current or most recent events and working backward in time. In addition, categories of events are grouped. Your work experience is separated from your education. Begin your work history with your present or most recent job and proceed to your earliest work experience. Most employers are interested primarily in your most recent experience, so more detailed descriptions are appropriate for the positions you held in the last five years. Beyond that, short summaries normally will suffice. Exceptions would be work or accomplishments that might be of special interest to prospective employers no matter when they occurred.

Always include dates with work experience, placed in a vertical column to the left of the experience description, on a line above the job title, or as part of the job description. An effective way to present experience that includes multiple assignments within a single organization is to show the total time spent with the organization to the left in a separate column and include the dates of the individual assignments in parentheses after the assignment descriptions.

Although education is also generally displayed in inverse chronological order, always list your highest degree first, even if it is not the most recent. Include dates of attendance, majors, and academic honors. If you include technical education or training sections, list the courses, seminars, and programs attended in inverse chronological order.

If you use a career summary, place it immediately preceding the detailed chronological work experience section. If you choose to use a job objective statement, insert it directly beneath your name and contact information. (See pages 112–113 for an example of a historical/chronological resume.)

Functional Resume

The functional resume is structured to present your professional skills or strengths under categorical headings without regard to dates or job assignments. Some headings typically found in a functional resume are Management, Research, Development, Design, Programming, Planning, or Projects. The headings or categories are up to you and can be as imaginative as you desire. In a functional resume, employers, dates of employment, and job titles are omitted in the functional descriptions.

By focusing the reader's attention on your skills, the functional resume can favorably present your background, even if you have had few jobs or limited experience, long periods of unemployment, or numerous jobs over a relatively short time. Although none of these circumstances may be your fault, employers generally have negative feelings about them and may make unfavorable assumptions about you when they see a functional resume. However, if you are an executive who has had a long career with only one or two companies, you can showcase specific skills and abilities very well in a functional resume. Likewise, if you are a new graduate or have limited experience, you can highlight personal traits and strengths.

The primary problem with the functional resume is the omission of dates. Surveys of corporate executives show that resumes without dates or with dates that are hard to find or cryptic raise suspicions that the candidate may be trying to hide something. Although this may not be the case, you will not be at the reading of your resume to offer an explanation. You may assume that if anyone "reads between the lines" of your resume, that reading will be negative. This is the only safe position you can take, and you should act accordingly in preparing your resume. One way to make this type of resume more effective and more credible is to include a brief, chronological listing of your work experience, including dates.

It is strongly recommended that you use the historical/ chronological format for your resume. Although you may have valid reasons for choosing a functional style, you should carefully consider its many pitfalls before choosing to use it. (An example of a functional resume appears on pages 174–175.)

Synopsis/Amplified Resume

The synopsis/amplified resume format includes a condensation of the pertinent data and an expanded work experience section that emphasizes specific areas. It is generally a two-page resume.

On the first page, list your identity items, then summarize your work experience, education, and personal information under a heading such as *Summary*. This is the "tell me about yourself in 100 (or 200 or 300) words or less" approach. It is also called a background summary or summary abstract.

On the second page, restate your employment history and expand significant areas. You can tailor this type of resume to emphasize areas of specific interest to a prospective employer. The synoptic/amplified format is also a good approach to use when your job duties cover more than your job title describes or when your work background is lengthy and varied.

The theory behind a synopsis description is that you are only going to get 20 seconds of the resume reviewer's time at most, so you should try to squeeze in as much information as possible in the hope of catching positive attention. An employer survey indicated, however, that employers generally were not impressed by the synopsis/amplified format, which they found repetitive and time-consuming to read. (An sample of a synoptic/amplified resume appears on pages 194–195.)

2

The Job Objective Statement

CAVEATS ABOUT THE JOB OBJECTIVE STATEMENT

The use of a job objective statement is optional. Including one in your resume may be helpful, harmful, or neutral depending upon the attitude of the person reading it and the way you write your objective.

If a prospective employer is looking to fill a position with an engineer who is seeking a position in research and development, and your statement says you are seeking a position in research and development, the statement has probably helped you. However, if the position you have applied for is not in research and development, the employer may look at your resume and your specific job objective and conclude that you would not be interested in the job offered. You, however, may be open to considering positions other than research and development. After all, you may have only put that objective in your resume because it was the ideal job; you actually would consider any job that is challenging and interesting. The problem is that the employer has already disqualified you on the basis of your job objective statement and you were not there to offer an explanation.

On the other hand, the job objective statement may not be considered by an employer one way or another. An enlightened employer may find your background and qualifications sufficiently interesting to want to consider you for an available position no matter what your job objective states. Such an employer believes that the position is challenging and can be sold to a qualified candidate. Such employers are rare. Most employers seeking to add to their staffs are very focused. If you say you want something that the employer does not offer, you will be believed and disqualified.

When you decide to send a resume to a potential employer, keep the following things in mind when writing a job objective statement:

- The prospective employer is a company you may want to work for, and you do not know what jobs might be available. There may be several positions in various areas you might qualify for.

- The job was described in a want-ad or advertisement. Although ads vary in the degree of detail and specifics offered about the job, they offer enough information to tell you whether your job objective statement will be helpful.

- If you have been requested to submit a resume, always ask, "Can you tell me something about the job?" The information you receive should tell you whether your job objective statement is appropriate.

- You are firm in what you want, have stated it clearly in your job objective statement, and do not wish to consider other opportunities. You will wait for what you want. In this case, screen jobs and employers carefully or you may waste a lot of money on stamps and envelopes.

- The resume must stand on its own. You will not be there to explain or refute anything in the resume. If the recipient of your resume misunderstands, misreads, or misinterprets it, you are stuck. Furthermore, you probably will never find out what the employer thought; you will be disqualified and that will be that.

If you decide to include a job objective statement in your resume, be informed about the job you are going after. If you prepare your resume on a word processor (and you should), it becomes a simple matter to change the wording of your statement to fit the job or delete the statement altogether.

DEVELOPING A JOB OBJECTIVE STATEMENT

If you decide to include a job objective statement in your resume, here are some tips to follow. If your resume is permanent, that is, typeset, you must be very careful and precise in your wording.

In your job objective statement, address either the job function (what you do), the job title (what you are called), or the job objective (where the job will take you). For example:

- Job function—"A position in a product design group."
- Job title—"A position as a Communications Engineer."
- Job objective—"A position leading to management of technical development projects."

Make your job objective statement concise, never longer than a few words or one sentence. Do not be too selective, however. A recent graduate entering the job market for the first time may have ambitions in specific areas, but would be happy to consider any position offering a reasonable career path. In fact, market conditions and the scarcity of jobs may make you even less choosy. For example, a statement such as "An entry level engineering position with a Fortune 500 company," limits you to Fortune 500 companies only. Better ways of stating the job objective might be "A position as an Electrical Engineer," or "An engineering position with a major international company." Even these statements, however, narrow your choices.

Another approach is to use a statement that combines a description of the job function or title you desire with a concise statement of your experience. For example, "A management position in Information Systems that will utilize the technical and management expertise I have gained in 15 years in the manufacturing industry."

Even this statement has set limits in several areas. Any job you would consider would have to be in management, and only in the area of Information Systems in a manufacturing environment. If the position required only 10 years of experience, you would probably be considered overqualified.

WHAT TO DO ABOUT THE JOB OBJECTIVE

Most job search books recommend using a job objective statement, which is viewed by many as standard on all resumes. As you can see, however, the problems it raises are significant. Once you launch a resume, it must stand on its own. You will not be there to say, "Yes, but I really meant …." If you choose to include a job objective statement in your resume, sanitize it so that it will cover almost any situation; or use a word processor so that you can tailor it to the specific job you are applying for. Another approach is to leave it out of your resume but include it in your cover letter. (See Chapter 9, The Cover Letter.)

3

The Career Summary

The career summary, or capsule resume, is a concise statement describing your qualifications, skills, and experiences in no more than two or three sentences or a series of short sentences. At 50 words or fewer, it is not easy to write. You must get to the point and not waste words, using action words and words that have the maximum impact. For example:

> "Over 15 years as Software Engineer and Systems Developer. Currently employed by Lockhaven Associates as Project Lead for upgrade on ADA compiler. Previously employed by Data Base Alliance Group and participated on three product development projects including design and development of flagship product, dBAG. Education includes MS in Computer Science and BS in Electrical Engineering. Holds two patents on software products."

A career summary is optional and may be left out of your resume. If you use a career summary, place it immediately following your name, address, and telephone number to get the greatest impact from the short reading you can expect your resume to get. If your career summary creates sufficient interest, the reader may continue with a detailed reading of the remainder of your resume.

Do not include a career summary *and* a job objective statement. Employers surveyed indicated that if one is used, the other should be omitted.

Notice the short sentence form employed throughout the summary, achieved primarily by eliminating "the," and "a," and personal pronouns. This style may not get high marks in grammar, but it is brief and to the point. Dates are also eliminated so that accomplishments from 10 years ago can be emphasized as well as more recent ones. Your goal is to create a good impression and gain the reader's interest. Include details about how long you have been employed at specific jobs, schools you attended, or additional details concerning your experience in the work history or experience sections of your resume. Remember, if your career summary does not get the reader's attention, the rest of your resume will not either.

The career summary is not for everyone. Use one only if you have enough experience and background to create an interesting capsule, generally, 10 or more years or if your experience is especially significant. As an entry-level candidate do not include a career summary—you have no career to summarize. If you are inexperienced, do not substitute statements about your personal characteristics or goals, which may read like a description of a Boy Scout: "Trustworthy, brave, clean, reverent, and generally a good all-round Scout." Although this example may seem farfetched, statements of motivation are equally unbelievable.

4

Experience

The work experience section is the most important section of your resume. Since it is the focal point of your presentation, you should place it in the upper third of the first page whenever possible.

In general, the information on your resume should appear as follows:

- Identification material
- Job objective or career summary
- Education
- Work experience

Depending on your level of experience, place the education section either after your work experience or before it, as listed above. As education is the most marketable asset for an entry-level person or a new graduate, place it in the most prominent position available near the top of the resume if you are a recent graduate.

Treat each job or period of employment as a separate item of information. List every company or organization that employed you and each assignment within each company as a section within that listing. Include starting and leaving dates in each listing; your job title, both official and generic; the official name of the company or organization; and the city and state where it is located. If you worked at a location other than the home office, list that city and state. Do not include the company's street address or zip code.

DATES OF EMPLOYMENT

List your dates of employment either in the left margin or at the beginning of the work experience description. Use a readily identifiable format. The following date formats are commonly used and accepted in the United States: July 10, 1994; Jul. 1994; July '94; Jul. '94; 7/10/94; or 7/94. Use only the month and year. Do not use military (10 July 1994), Julian (4182), or European (94.10.4) formats, which can be misinterpreted or can prejudice the reader.

JOB TITLE

The job title should be the first item in each work experience description. If the job title is descriptive of the work you performed, it may stand alone. For example, if your job was Senior Design Engineer, this title describes your job sufficiently. However, some companies have in-house or organizational titles that are not descriptive. One aerospace company uses titles such as Member of the Technical Staff. These titles may be recognized within the organization, but they are meaningless to anyone on the outside. If you have such a title, list the descriptive title first, then the official title in parentheses. For example, Development Project Team Member (Member of the Technical Staff).

JOB DUTIES

The description of the duties your job entails is the most important section of your resume, and you will spend the bulk of your time developing it. Providing an accurate and complete description of your work experience and assignments while limiting what you say to as few words as possible is a formidable task. However, by following a few simple rules, you will be able to describe your job duties clearly and concisely and increase the effectiveness of your resume.

Rule 1. Be Brief

Limit each job description to approximately seven or eight lines if possible. If you must exceed the eight lines to tell your story, don't worry. Just remember that brevity is appreciated by your reader. Make every word count and carry as much impact as possible. Use the "Action Words" guide on pages 14 and 15 to find dynamic verbs that project you with as much punch as possible.

Your current or most recent job is also the most relevant one. Make its description the most detailed and longest. Apply the "seven or eight line" rule to jobs you held up to 5 years ago. Descriptions of jobs held between 5 and 10 years ago should occupy only half to two-thirds of that space, four to five lines. Beyond 10 years, use a one to three line summary.

One way to be brief is to exclude adjectives, adverbs, and pronouns. This will not win you any writing contests, but it keeps your description to the point and projects the message you want to send. For example, instead of the first person, "I was responsible for the design and development of ..." write, "Responsible for design and development of ...." The resume is obviously about you, so using the pronoun "I" is unnecessary.

Do not use the second person, "He was responsible ...," which makes the resume appear to be about someone else and sounds strange and distant. You want to avoid the possibility that the reader might assume that the resume was written by someone else.

Rule 2. Keep Verb Tenses Consistent

Use present tense to describe your duties if you are currently employed, and past tense for previous assignments.

- Present tense: "Implements design changes ...."
- Past tense: "Implemented design changes ...."

Rule 3. Use Active or Power Verbs

Active verbs such as "designed," "implemented," "programmed," "planned," and "performed" add power and energy to your resume and conjure up descriptive images in the mind of the reader. They give your resume authority. Use them whenever possible.

ACTION WORDS

Regardless of how your resume looks, without the right words to convey your abilities and skills it is empty form. No one but you can adequately describe your background and experience, but by using certain words in your descriptions your strengths and abilities will be better represented. Use these words when you are preparing your Employment History Worksheet (Chapter 9).

A accomplish, achieve, account, accumulate, acquire, activate, adhere, administer, advertise, advise, allocate, analyze, appraise, approve, arrange, assign, assist, assume, assure, audit, augment, authorize, automate

B brought, budget, built

C   catalogue, change, code, collect, communicate, compare, compile, complete, compose, compute, conceive, concentrate, conduct, configure, consider, construct, consult, continue, contract, contribute, control, cooperate, coordinate, correct, correlate, create, credit

D   debut, decrease, define, delegate, delete, design, determine, develop, direct, disperse, display, distribute, document

E   edit, educate, emphasize, employ, engage, engineer, enhance, enlarge, ensure, establish, examine, execute, exercise, expand, expedite, extend, evaluate

F   fix, flowchart, forecast, functioned as, furnish

G   generate, grant, graph, guarantee

H   head, help, hire

I   implement, improve, include, increase, inform, initialize, initiate, innovate, inspect, install, instruct, integrate, interface, interpret, interview, introduce, install, invent, investigate, involve, issue

J   join, justify

L   learn, lease, led, lessen, load

M   maintain, manage, market, master, measure, meet, modify, monitor, motivate

N   negotiate, neutralize, normalize, notify

O   open, operate, orchestrate, organize, order

P   participate, perform, persuade, plan, post, prepare, present, process, procure, produce, program, project, promote, propose, protect, provide, publicize, purchase

Q   qualify, quantify

R   reclaim, recommend, reconstruct, recruit, release, report, represent, request, require, requisition, research, resequence, reshape, responsible for, retrain, retrieve, review, revise

S   save, schedule, screen, secure, select, sell, serve, set objectives, set up, solve, sort, specify, staff, standardize, stimulate, strengthen, structure, subcontract, submit, succeed, summarize, supervise, supply, support, synthesize, systematize

T   teach, test, trace, track, train, transfer, translate

U   update, upgrade, underscore, utilize

V   validate, verify, visualize

W   write

Rule 4. Use Short, Direct Phrases

Using short sentences with strong action verbs and phrases makes your resume appear more dynamic, and because the resume represents you, you appear more dynamic. Long, rambling sentences connected with numerous "ands" may confuse your reader and promote misreading and misunderstanding.

Rule 5. Use Bullets

Rather than lumping your accomplishments in a single paragraph, list distinct job responsibilities for each employer or assignment and highlight them with a bullet or distinct symbol. Highlighting your accomplishments makes the information easy to read and digest. This technique is well accepted and even preferred by many employers.

Following are some of the more creative symbols available for bullets: ❑ ◆ ● ❖ ☞ ✔ ✐

EXPERIENCE:

Aug '88 to Sep '94	**Project Engineer**
	❑ Designed/developed HF receiver
	❑ Prepared PERT and long-lead programs
	❑ Designed and developed microwave systems and antenna equipment for EW application in ECCM

Rule 6. Be Accurate

A technical resume should be accurate in the extreme. If you make a mistake in the model number or use of technical equipment or issues, the reader will assume that you do not know what you are talking about. You must refer to the tools and equipment of your trade correctly.

Rule 7. Place Information Strategically

Normally it makes little difference whether you list your *work experience* or *education* first on your resume, or, within your work experience section, you list the *company name* or a job title first. Sometimes, however, these considerations make a difference. For example, when a prospective employer has specified that candidates be degreed, display that information prominently near the beginning of your resume. Even the smallest detail can make the difference between getting an interview and being rejected out of hand. Remember, you can take few single actions with your resume that will guarantee success. However, by giving careful deliberation to each issue and making the choices that may have even a minimal effect, you will have a better overall resume and improve your chances for success.

5

Education

Your education is an important item and it belongs on your resume, with two exceptions.

Do not list your education on your resume if it might arouse a bias or prejudice in the reader. For example, if you attended or graduated from a religious institution or foreign university and are concerned that an employer might be biased against such a background, leave out your education. Consider carefully, however, whether you would want to work for a person who would so judge you.

Leave out your education as well if the employer has specified a degree or a major that you do not have. Perhaps your experience would qualify you for the job. Focus your reader's attention on your experience by omitting your education. The employer will not know whether you had the education or not and therefore may be forced to judge you solely on your experience. The employer might disqualify you in any case. However, if you included your education and it was inadequate, you would almost certainly be disqualified. Without it, your chances of being called in for an interview would improve.

If you have college credit but do not have a degree, list your education as years of college. For example, if you lack one semester's credits for your degree, write "State University, 3 1/2 years . . . ." If you will receive your degree in the near future, you may list the date. If you have accumulated more credits than required for graduation but they are not the right credits, do not list more than 3 1/2 years of college. Stating that you have five years of college but no degree may cause the employer to wonder why.

Writing a resume is a lot like playing a game. It would be a tremendous advantage if you knew what an employer was thinking before you constructed your resume, but you do not. The most you can hope or work for is an educated guess.

PLACEMENT

As mentioned, the placement of your education on your resume may be an important issue. If your education is recent or your training is pertinent to the position, list your education first, immediately following your name, address, and telephone number. For example, if you hold degrees in engineering and the job you are applying for is engineer, you want your education to stand out. If you received your degree more than 10 years ago, or if it is unrelated to the job or your experience, such as a degree in the liberal arts, place it after your work experience. The primary criterion for placement of any information on your resume is that items of primary importance to the job you are applying for be placed where they will be seen first.

ORDER

Apply the rules of inverse chronological order or level of degree to your education. List your most advanced degree or your most recent among equal degrees first. For example, a Ph.D. would appear first, then an M.S. or M.A., and, finally, a B.S. or B.A. degree.

INFORMATION TO INCLUDE

Include in each degree you list: the degree; the major or designation (e.g., Chemistry, Electrical Engineering, Computer Science); the name of the college or university; and the year the degree was awarded. Do not include the month. Do not include the year the degree was awarded if you received it more than 10 years ago. In the technical world, a degree older than 10 years can be perceived as out of date. Although you may have kept your knowledge current, your prospective employer might conclude that your knowledge, like your degree, is out of date.

OPTIONAL ITEMS

The following are optional, to be included on your resume only if they might favorably influence prospective employers.

- The address of the college/university attended. Include only the city and state, not the street address.
- Majors and minors. Include both if you are a recent graduate with limited work experience.
- Honors, awards, scholarships. If you are a recent graduate, include honors, awards, or scholarships that you may have received while getting your education. Mention your grade point average (GPA) if it is 3.0 out of 4.0 or higher, and any scholastic honor societies you belong to.
- Extracurricular activities. Include extracurricular activities if you are a recent graduate. Emphasize organizations such as honor societies and professional groups that will give you a professional image; play down those that are social or play oriented, such as social fraternities and sororities, sport teams, and hobby groups. By including nonscholastic groups, you may appear well rounded. However, consider who will be reading your resume and the message you want to send.
- Technical and professional education. Include trade schools, seminars, and military training in a separate section labeled "Technical" or "Professional Education" following your "Work Experience."
- Personal data. In the not-too-distant past, most resumes contained information concerning age, height, weight, marital status, number of children, state of health, hobbies, and other interests. Although many resume preparation books and manuals advise you to include this material in some form, do not do so. Most recruiters feel it serves no valid purpose and may even lessen the impact of your resume. There are no rules about volunteering personal information. However, because there are laws that address what information an employer may or may not legally inquire about, some employers will not consider resumes that contain personal data.

Federal, state, and local statutes prohibit employers from discriminating on the basis of age, sex, marital status, race, national origin, religion, medical condition, or sexual orientation. Something so innocuous as asking a person's age may seem unimportant, but if that person is over 40 years of age, he or she is protected from age discrimination. Most employers are smart enough not to open themselves to legal action by asking questions that concern the issues covered by the various laws. If job candidates openly place their ages on resumes, an employer who rejects them might be open to challenge. Hence, an employer may conclude that it is better not to consider such resumes. Prospective employers can deduce candidates' ages from their graduation dates—another reason not to include them if you graduated 10 years ago or more—and in an interview. Employers who discriminate on the basis of age—and they do—can do so through other methods.

Height and weight are unimportant unless you are applying for a position that has minimum physical requirements, such as the police or military. You want the employer to focus on your qualifications for the job, not what you look like.

Health is an issue about which you should volunteer nothing. If you mention it, do so only because your health is excellent. Although you may have a health problem or a disability, you must be judged on your ability to perform at a particular job and not on physical criteria. Do not give an employer a reason or opportunity to disqualify you, however, before an interview. Get the interview, demonstrate your qualifications, and let the employer deal with your disabilities face to face. Hires are made primarily on the basis of perception and emotional criteria. Your goal is to get in front of the employer and offer convincing arguments and proof that you are the right person for the job. Don't let your resume do you a disservice.

Marital status and number of children should also be left off the resume. It was once thought that married *men* with children made the most stable employees. That information became an asset on the resume. On the other hand, a married *woman* with children is still likely to be perceived as a risky hire because of her family obligations. A single woman may be considered risky because many employers may believe that she'll quit and get married. Single also may mean something other than "never-been-married"; it may mean "divorced," "separated," or "gay." These stereotypes can bring you grief if you give an employer the opportunity to judge you. When you place something on your resume and send it out, you run the risk of misinterpretation or speculation by whoever reads it. Placing the number of children or their gender and ages on the resume is generally done to create an image of family stability. Because you don't know what the person reviewing your resume is going to infer from such information, it is better to exclude a "Personal Information" section altogether. Your personal life and lifestyle are nobody's business but yours. The only concern that an employer should have is that you will show up for work and perform satisfactorily.

Hobbies and outside interests may be noted in a separate section if including them will add to your marketability. A word of caution— although your motivation in providing this information is to show that you are an all-around, balanced person, you do not know what inferences a prospective employer might make based on this information. If your research on the employer turns up information that you share similar interests, listing your hobbies and interests could have a positive effect. In general, you would be fairly safe in listing any career-oriented hobbies or interests, such as computers or job-related workshop projects.

HONORS, PUBLICATIONS, MEMBERSHIPS

- Licenses, clearances, or special qualifications. Include these in a special section. Accomplishments rewarded and recognized by others lend credibility to your professional status.
- Books, papers, publications, and presentations. List publications and presentations to demonstrate capabilities above and beyond your educational credentials. Employers who place a high value on degrees and other educational accomplishments tend to be more impressed by publications than employers who are interested primarily in your work experience.
- Professional societies and organizations. List your memberships in professional organizations under a heading of that name. Indicate whether you are currently serving as an officer or have been an officer.

6

References

You may end your resume with a statement that references are available on request, although this statement is optional. Obviously, if an employer wants to contact references, he or she will ask for them. A general rule in preparing resumes is to eliminate extraneous or unnecessary items. The statement that references are available on request is extraneous. It is included more out of custom or general usage than necessity.

FORMAT FOR A REFERENCE STATEMENT

If you decide to include a Reference statement, keep it simple and straightforward. Here are some acceptable examples:

References: Upon request
References: Will be furnished on request
References: Provided on request
References provided upon request.

Do not editorialize with declarations such as "Excellent references available on request." Would you provide less than excellent or unfavorable references? Listing personal or professional references on a resume is unprofessional and may result in unnecessary bother to those you have listed. Never list specific references on your resume.

USING REFERENCES CORRECTLY

Preparing a list of references is a necessary part of resume preparation. Although you will not send it out with your resume, you should have it ready for employers who request it. The fact that they ask for it is a positive sign and an opportunity for you to interact and establish positive rapport.

Begin preparing your reference list by making a preliminary inventory of the persons you might wish to include.

Professional and Personal References

Generally, plan to have a list of three professional and three personal references. Professional references should be persons who know your work and your professional qualifications. The best are recent supervisors and managers who you reported to or who know your work. Less desirable is listing co-workers. A word of caution—listing only co-workers and no managers may cause a prospective employer to wonder whether no manager would recommend you and you had to rely solely on your friends. Personal references are less important than professional references. Most employers are primarily interested in your work performance. Those who check personal references are most often checking your character or financial stability.

Getting Permission

Before placing a reference on your list, get permission from the person to do so. Ask prospective references whether you may list them as a reference and tell them something about your job or career objectives, so they can speak knowledgeably and favorably about you. Listen carefully to their responses when you ask to list them. If they seem reluctant or indicate any negative issues, do not include them. You want references who will be enthusiastic about you and give you a positive recommendation.

Secondary References

In your conversation with prospective references, ask who they would refer the employer to if they were asked about someone else who might know about you. This secondary reference can be even more significant than the primary reference. An employer will assume you will list only those people who will recommend you favorably. A secondary reference adds depth to your qualifications because it does not come from you but rather from your primary reference. You should also contact secondary references and mention that they may be contacted. Evaluate them with the same measure as you did your primary references.

Information to Give Your References

Discuss your work and personal history with your references since your last contact with them as well as changes of address and phone numbers, and other personal data such as a name change through marriage.

Reference Contact Information

Provide your prospective employer with adequate contact information for all references. Include their work and home telephone numbers, current work addresses, and the times that they can be reached. Also provide their current titles and employers and their titles and employers when you worked with them. Here is an example:

Mr. Bill Charleston, Director of R&D
Acme Laboratories, 800 Research Road, San Jose, CA 91222
Work: (415) 555-3232   Home: (415) 555-0987
Contact at work 9 AM to 5 PM. At home, after 7 PM, before 10 PM.
Project Supervisor at Jones Electronics, 1991–93.

7

Information Never to Include

A resume is not a biography or a historical summary of your life and career. You are not required to lay out your darkest secrets for the world to see. A resume is a marketing device that should be structured to present your professional and work background from the most favorable perspective—nothing else. Most people would not intentionally include negative information in their resumes. However, often information is included in a resume with good intentions that can bring negative reactions.

SALARY ISSUES

Never give your current salary or past salaries in your resume, despite demands in want-ads that you provide your salary history and will not be considered if you do not. Prospective employers ask for a salary history to disqualify people. If your current salary is higher than a prospective employer is willing to pay, you're out! If your current salary is lower than the employer expects to pay for the position for which you are applying, the employer may assume that you do not have enough experience for the job being advertised or you are worth only what you are being paid. Your salary history may fail to follow the progression that whoever reviews your resume thinks it should. If so, you are disqualified. You do not know what the expectations of the resume reader are, and you cannot alter your salary history in any case. Simply omit any mention of salary and concentrate on establishing your qualifications for the position.

In addition, even though employers routinely ask for a salary history, most feel that a person's compensation is a private matter and a confidential subject. In many companies, discussion of salaries is forbidden, and so no mention of it should appear on your resume.

Your resume may also pass through several hands before it reaches the decision maker who has the power to invite you in for a face-to-face interview. How much you are making now or may make is no one's business. Including that information on your resume may give cause to discard your resume before it reaches the right person. Many qualified candidates have done end-runs around administrative functionaries and made contact with the hiring manager only to find that their resume had been discarded before the hiring manager had seen it. Do not give anyone a reason to toss your resume!

Salary can be a sensitive issue with potential for causing problems. It is best left for your face-to-face interview or your cover letter. (See Chapter 10, The Cover Letter, for more information about salary.)

If you are overcome with the need to put something on your resume that refers to salary, the only acceptable comment is "Salary is open."

REASONS FOR LEAVING

Your resume is definitely not the place to discuss why you left a previous employer, even if your reasons were positive. Allowing any speculation by the reader of your resume about reasons for leaving a job is dangerous unless you are there to discuss it.

NAMES OF CHILDREN AND SPOUSE

Although this has been discussed previously, it needs to be said again. Do not include any information about your spouse or children. Your personal life has nothing to do with your qualifications for the job, and information about it does not belong on your resume.

PHOTOGRAPHS

Never send a photograph with your resume. It is questionable and possibly illegal for an employer to request a photograph unless you are an actor or a model. A photograph can identify your race, age, sex, or appearance. This information can lead to discrimination against you. If you feel that you are an attractive person and believe that a photograph will help you get the job, forget it. Wait until you meet the employer and then present the real you. Sending a photograph may also send the wrong signals to an employer with less than honorable intentions, a problem you do not need.

8

The Attention-Getting Resume

The job of your resume is to get you interviews with people who have the power to hire you. To do so, resumes must catch their attention positively and create enough interest for them to invite you to face-to face meetings.

If your resume is difficult to read, too long, poorly organized, unattractively displayed, or has an unprofessional appearance, it most likely will be ignored and discarded. If it does not get the reader's attention immediately, it isn't going to be read and you are not going to get an interview. In fact, an unattractive resume very likely will be discarded immediately—especially if it must compete with good-looking, well-constructed competitors. On the other hand, if your resume is thoughtfully constructed, with an uncomplicated format, good use of margins, and easy-to-read text, it probably will be selected for further examination.

Clearly, however, the appearance of your resume will not overcome a lack of substance in its content. A professional-appearing resume will attract attention, but it must also have something good to say to be successful in getting you interviews.

PROOFREAD YOUR RESUME!

A very important step in creating an attractive resume is to proofread it. A resume with misspelled words, grammatical mistakes, or typographical errors not only says that you are a person who does sloppy work but also publicly represents you that way. The message is that you will represent an employer the same way. Your resume will end up in the "round file"—and rightfully so!

Proofread your resume at least twice, then have a friend proofread it. This second proofreader should be a real friend who will offer hard, constructive criticism and tell you whether your resume would create enough interest for the reader to call you in for an interview. Use the spelling checkers on your word processing program as well.

USE HIGH-QUALITY PAPER

The quality of paper you use for your resume may appear to be of little significance. However, like other aspects of resumes, if there is only a fraction of a percentage point of a chance that paper might have some impact, consider it.

Type of Paper

Use a high-quality 20- to 24-pound bond paper with a cotton rag content of 25 percent. It is slightly heavier than regular typing or copier paper, feels good to the touch, and will stand up to handling by a number of readers. Avoid using erasable bond, onion skin, computer, or cheap copier paper. If you use a computer printer with continuous-form paper, use a high-quality paper with laser cuts on the form feed portions to eliminate fuzzy edges. Never use company letterhead or company copy machines that put the company name on the edge of copies.

Size of Paper

Produce your resume on standard 8 1/2- by 11-inch paper. Folders, files, and office equipment are made for the standard paper size, so your resume will always fit. Odd shapes and sizes of paper cause problems with filing and often are discarded. Being different may draw attention to your resume, but that attention must always be positive. Fold-over resumes, brochures, and bound documents are less acceptable than the standard size.

Staying within size limitations or other parameters should not discourage you from considering bold or daring resume designs. However, you should carefully consider the consequences of such actions.

LIMIT THE NUMBER OF PAGES

Although the standard advice is to try to get your resume on one page and never more than two, this is often easier said than done. Getting 20 years of experience on one or two pages is as difficult as trying to fill one page with no years of experience. A more practical rule is to aim for one or two pages, but if your resume requires more pages to be complete, use more. Always try for as few pages as possible. The key to getting a longer resume read is making what you say in the first few lines interesting. You only have the reader's attention for a few seconds. Make them count.

If your resume is more than one page long always put your name at the top of the additional pages. Print on one side only and staple all the pages together in the upper left-hand corner.

You can reduce the length of your resume and maintain its content by eliminating pronouns, truncating sentences, and formatting pages. Do not reduce the type size to fit your resume on one or two pages. The larger the type, the easier it will be to read and the more likely it will be read. Here is a sample of 4-point type. Would you like to receive a resume like this?

LIMIT THE NUMBER OF PAGES

Although the standard advice is to try to get your resume on one page and never more than two, this is often easier said than done. Getting 20 years of experience on one or two pages is as difficult as trying to fill one page with no years of experience. A more practical rule is to aim for one or two pages but if the your resume requires more pages to be complete, use more. Always try for as few pages as possible. The key to getting a longer resume read is making what you say in the first few lines interesting. You only have the reader's attention for a few seconds. Make them count.

If your resume of more than one page long always put your name at the top of the additional pages. Print on one side only and staple all the pages together in the upper left-hand corner.

You can reduce the length of your resume and maintain its content by eliminating pronouns, truncating sentences, and formatting pages. Do not reduce the type size to fit your resume on one or two pages. The larger the type, the easier it will be to read and the more likely it will be read. Here is a sample of 4-point type. Would you like to receive a resume like this?

GO EASY ON COLORS

Be conservative in your choice of paper and ink colors. Use black ink and white, off-white, cream, light gray, or buff paper. Striking colors may get you attention, but that attention is likely to be negative. Although many businesses seek creativity and flair, they are not likely to appreciate it in your resume. Evaluate your prospective employers carefully before deviating from the accepted inks and papers.

PAY ATTENTION TO WHITE SPACE AND LAYOUT

White space is the area on your resume not filled with print, such as margins and the spaces between paragraphs. Use the margins around your text for maximum eye appeal and effectiveness. Center your text with 1 to 1 1/2-inch margins at the top and bottom and 1-inch margins on the sides.

Between blocks of information, leave several lines of space to keep the reader from being overwhelmed. Do not squeeze information in with minimum or no spacing in order to keep your resume at one or two pages. Keep your information brief. Likewise, do not push the margins out to fit in more information. Both of these tactics reduce the eye appeal of your resume and make it more difficult to read.

REPRODUCING YOUR RESUME

Word Processing

In Chapter 14, Sample Resumes, several resumes are shown in two formats, the standard single type font and single point size resume that has been around for years and an enhanced resume created using the word processing, desktop publishing, and graphics capability of personal computers.

Selecting a Personal Computer

The term *personal computer* carries a much larger connotation than merely a class of computers. PC users become very attached to their computers and are almost fanatical in their support of specific software packages or hardware. The users of Macintosh and IBM PCs and various clones defend their choices vigorously.

Which System Is Best?

The truth is that there is no "best" hardware or software system. Both Macintosh and IBM and its clones are excellent computer systems. Base your choice of a computer on your needs. If you are a first-time buyer, talk to friends who own computers, listen to their recommendations and complaints, visit various dealers, try out different systems, assess the applications you plan to use, and make an informed choice.

Word Processoring Software

The most popular use for the PC is word processing. Word processing is the creation of text using operations normally associated with a typewriter. However, word processing is much more than typing. Word processing permits character and word replacement, typeovers, inserts, moving whole blocks of text, document formatting, changing fonts and scaling, and preview prior to printing. In addition, many popular word processors offer features such as What You See Is What You Get (WYSIWYG) so that the display on the screen is the same as what will be printed, spell checking, and grammar checking.

The ability to produce printshop-quality documents and restructure them at will is sufficient reason for job hunters to toss their typewriters away and embrace word processing. However, word processing offers much more.

The manuscript for this book was prepared using Microsoft's *Word for Windows 2.0*. It is one of the major word processing packages and offers all the features previously mentioned. Other popular word processing packages include *WordPerfect 5.2* and *WordPerfect for Windows, AMIPro,* and *WordStar 7.0*. All of these are excellent choices for resume preparation and generation. The individual cost of these word processors is about $400. There are cheaper packages with fewer features that may also be suitable. These include *PFS-Write* as well as scaled-down versions of the major packages.

A major reason for using a PC with word processing is the ability to employ a variety of typefaces (fonts) and reformat your documents at will. The manuscript for this book and the sample resumes were prepared in the Truetype fonts that are standard with *Microsoft Word for Windows 2.0*. An add-on software package with additional fonts was also used. This package is called *MoreFonts 3.0* by Micro-Logic Software, and it permits the addition of 28 scalable typefaces. The software also allows the addition of patterns, outlines and shadows, and backgrounds to the basic typefaces which can be combined to give an almost infinite variety of fonts. There are also many other add-on software font packages available, and the price is very reasonable. *MoreFonts*, for example, retails for $99.95.

Selecting a High-Quality Printer

Before you consider the software to help you with your job search, there is one piece of hardware besides the computer that you will need—the printer. If you are going to produce high-quality resumes, you will require a printer capable of producing print-shop quality output. There are two choices that will provide that level of quality. The first is the laser printer and the second is the inkjet printer.

Laser printers are available in desktop models from smaller personal models to office models, ranging in price from around $700 to $1,800, depending on speed and features desired.

The second choice is the inkjet printer, a cheaper choice with prices ranging from around $400 to $700. Inkjet printers operate by spraying ink through a small nozzle onto the paper. The quality is similar to that achieved by laser printers, but they are slower.

A dot-matrix printer offers the most economical method of computer printing. If you opt for this choice, use a 24-pin model only. Nine-pin models are available for less money, but you will not get the letter-quality printing you need for a good-looking resume.

Typesetting and Offset Printing

In the past, the best-looking resumes were those that were professionally typeset and printed. The advantages of using typesetting have been the availability of a large number of typefaces and fonts; a professional printer and typesetter to advise you; and a crisp, clean output.

The major disadvantage of a typeset resume is that once it is produced, you must live with it unless you want to go through the process and expense again. However, one resume format and layout will seldom fit every situation or prospective employer's needs. Even a great-looking resume will fall flat if it doesn't say the right words in the right way for a specific job.

Although using a typeset resume will guarantee you a rich-looking, high-quality product, there are acceptable alternatives that will serve you equally as well and give you greater flexibility in the use of your resume. If you own or have access to a word processor, you will be able to change your resume to target it to specific job opportunities. Not all word processing systems' printers offer the quality of reproduction that you need to get your resume read, however.

Typewriting

In this day of low-cost personal computers and inexpensive, full-featured word processors, using a typewriter to produce your resume should be a last resort. If you do use a typewriter, however, it should be an electric, commercial model with self-correcting features. Some models are equipped with built-in spelling checkers and display screens. These stand-alone word processors are relatively inexpensive.

Many models also come with interchangeable "daisywheels," "golf balls," or "thimbles," which allow you to intermingle type sizes, include regular, boldface, and italic fonts, and create special effects with unique fonts.

Personal computers have reached a price level where models can be purchased for less than the cost of a commercial-model office typewriter. Finally, most jobs, especially technical positions, require some degree of computer literacy. A non-computer-produced resume is a dead giveaway that you lack these skills.

Photocopying

Once you have either a typeset or laser-printed resume, photocopying is the least expensive method of reproducing several copies. Make sure the copies are free of imperfections, with good contrast. If your resume is neat, attractive, and well constructed, it will be acceptable.

9

Employment History Worksheet

The following worksheets will assist you in organizing your background information and developing your resume. These worksheets correspond with standard resume formats. When completed, your worksheet will be a skeleton version and a working first draft of your resume.

FOR EXPERIENCED TECHNICAL PROFESSIONALS

Identifying Material

Name: _____

(Use full name or your name as you sign your checks. If you have changed your name during the work history recorded on your resume, you may include any name by which you may be known.)

Address: _____

(Street & number, apartment number, city, state, zip code)

Home telephone: _____

(Include area code)

Business (office) telephone: _____

(Include area code. You may also wish to include the times you are available.)

Job Objective

Job objective is optional. Target it to the job you are going after.

Job objective: _____

_____

_____

_____

Resume Summary

The resume summary, like the job objective, is optional. Keep it brief and focus on the specific job for which you are submitting your resume.

Resume summary: _____

_____

_____

_____

Clearances

Clearances include active or inactive security clearances, security and financial bonds, and pertinent background investigations.

Clearances: _____

Employment History

List employment history in inverse chronological order, that is, beginning with your current or most recent job.

Dates   *Description of Responsibilities*

From/to
(Mo/yr to mo/yr)

_____   _____

                        _____

                        _____

                        _____

Name of company: _____

Address of company: _____
(City and state only)

Job title: _____
(Use your official job title if it is descriptive of the position. Otherwise, use a generic title.)

Dates   *Description of Responsibilities*

From/to
(Mo/yr to mo/yr)

_____   _____

                       _____

                       _____

                       _____

Name of company: _____

Address of company: _____
(City and state only)

Job title: _____
(Use your official job title if it is descriptive of the position. Otherwise, use a generic title.)

Dates   *Description of Responsibilities*

From/to
(Mo/yr to mo/yr)

_____   _____

                       _____

                       _____

                       _____

Name of company: _____

Address of company: _____
(City and state only)

Job title: _____

(Use your official job title if it is descriptive of the position. Otherwise, use a generic title.)

Dates	*Description of Responsibilities*
From/to	
(Mo/yr to mo/yr)	

_____   _____

_____

_____

_____

Name of company: _____

Address of company: _____
(City and state only)

Job title: _____

(Use your official job title if it is descriptive of the position. Otherwise, use a generic title.)

Education

List your education in inverse chronological order as you did your employment history. List your most advanced degree or your most recent educational experience first.

Degrees or Attendance at Colleges or Universities

Dates   University: _____

From/to: Address: _____
(Yr/yr or   (City and state only)
graduation
year only)

_____   Degree: _____

              Honors: _____

Technical Education or Training

Dates   Course: _____

(Mo/yr   Institution: _____
graduated
length of course)

                     Location: _____
                     (City and state only)

Personal Information

Personal information should be included on your resume only for a compelling reason. The following information may be included.

Willing to relocate: _____

Willing to travel: _____

Hobbies or interests: _____

Professional Memberships and Affiliations

_____

_____

_____

Publications, Speeches, and Special Achievements

_____

_____

_____

Foreign Languages and Special Skills

_____

_____

_____

_____

REFERENCES

Never include the names of your references on your resume. However, have them available if they are requested by a prospective employer. Have a minimum of three professional references, former employers, supervisors, or fellow employees; and three personal references, persons who can attest to your character.

Name of reference: _____
(Provide first and last name and nickname, if used.)

Position: _____
(Provide current job title and job title at the time of your association. Also include company at the time of association if reference has moved.)

Current company/company at time of affiliation: _____

_____

Company address: _____
(Provide complete address of current company, street address, city, state, and zip code.)

Home address: _____
(Provide if known. Include street address, city, state, and zip code.)

Business telephone/extension: _____
(Include area code)

Home telephone: _____
(Include area code and availability, if known.)

Name of reference: _____
(Provide first and last name and nickname, if used.)

Position: _____
(Provide current job title and job title at the time of your association. Also include company at the time of association if reference has moved.)

Current company/company at time of affiliation: _____

_____

Company address: _____
(Provide complete address of current company, street address, city, state, and zip code.)

Home address: _____
(Provide if known. Include street address, city, state, and zip code.)

Business telephone/extension: _____
(Include area code)

Home telephone: _____
(Include area code and availability, if known.)

Name of reference: _____
(Provide first and last name and nickname, if used.)

Position: _____
(Provide current job title and job title at the time of your association. Also include company at the time of association if reference has moved.)

Current company/company at time of affiliation: _____
_____

Company address: _____
(Provide complete address of current company, street address, city, state, and zip code.)

Home address: _____
(Provide if known. Include street address, city, state, and zip code.)

Business telephone/extension: _____
(Include area code)

Home telephone: _____
(Include area code and availability, if known.)

10

Cover Letter

As important as your resume is in helping you get a face-to-face interview with a prospective employer, never send it by itself. Each time you mail your resume, you must include a cover letter.

The cover letter is a formal letter that introduces you and your resume to a prospective employer. An act of courtesy and a necessary piece of business, the cover letter indicates that you are a serious and professional job seeker. Regardless of whether it is addressed to a specific individual, a job title, or a box number, it is a *personal* letter. It tells the recipient that you value this contact and have made a special effort to consider the specific requirements of the position. A cover letter addresses your specific desires and emphasizes the skills and talents that qualify you for the position.

Although the cover letter increases the impact of your resume, it is not intended to stand in place of the resume. An excellent cover letter accompanied by a poorly prepared resume will not serve you well. The two are complementary and mutually supporting. Both the cover letter and resume introduce you, and both must proclaim excellence.

Brevity is the key to the cover letter. Make it one page, containing no more than four paragraphs. A short, to-the-point letter is more likely to be read. Keep in mind that your goal is to get the reader to consider the skills and experience described in your resume and that the cover letter should tantalize the reader with two or three attention-getting paragraphs. Finally, four or fewer paragraphs will place a limit on what facts you can include. Too much information makes discouraging reading.

Always try to address your cover letter to an individual by *name*. Do not address it to a title only unless you have exhausted all efforts to discover the name of the person who belongs to the title. Usually you can get the right name through a phone call to the company and a direct question, without identifying yourself. Be sure to get the correct spelling of the person's name; his or her preferred form of address (Mr., Dr., Ms, Miss, Mrs.); and the precise job title: "Engineering Director" is not the same as "Director of Engineering." You may also find a name to go with a title through a business directory such as Standard and Poor's, the Fortune 500 (or Double 500) Directory, or other national or local sources available at the public library. Calling the company is preferable, however, because the directory may be out of date. When answering a newspaper or Box No. ad (also known as a "blind ad"), you may address your cover letter to whomever or whatever title is included in the ad.

In general, executives and managers do not receive their mail directly. Many have their secretaries sort and organize their mail by an established priority system: first, mail of an urgent business nature, then personal mail, then mail of unknown value addressed to the person's name, then junk mail. Your hope is that the position you are applying for is of enough importance that your mail will be at the top of the pile. However, it is also likely that you will fall into the "unknown value" category. Obviously, you *never* want to be included in the junk mail.

Just as you used a good-quality paper with your resume, you should also use a high-quality white or near white, 8 1/2- by 11-inch paper for your cover letter. Personal stationery with your name, address, and telephone number is appropriate to use if you have it.

ORGANIZATION OF THE COVER LETTER

Heading

The heading includes the date of your letter, the name and title of the person you are writing to, and the mailing address, including zip code. The date on the letter should be as close to the day you mail the letter and your resume as possible. A stale date looks as if you delayed mailing for some reason, and a postdated letter that arrives before the date makes you look as if you don't know what you are doing.

Begin the mailing address with the name of the person you are trying to reach, using the information you gathered in your telephone call to the company. If you were unable to get information about either the title or the gender of the person, use the name by itself with no title. Put the title of the person to whom the letter is addressed on the next line. The street address, including street and number, suite or office number, and mail stop or mail code, follows on the next line. Finally, the last line of the address contains the city, two-letter state abbreviation in capital letters, and zip code. Use the 13-digit code, if you know it.

Salutation

The salutation, or greeting, sets the tone of the correspondence. For the cover letter, it should begin "Dear [Mr./Ms. _____]" addressing the person by his or her title and last name. A "first name" greeting is permissible only if the person is an acquaintance or friend; otherwise it is too familiar.

Body

Four paragraphs or fewer should suffice for an effective cover letter. You want to whet the appetite of the reader to read your resume to see what this interesting person is all about.

First Paragraph

The first paragraph is a straightforward statement of why you are writing to this particular employer. If you are answering an ad, give the name and date of the publication where the ad appeared. If you are writing on the recommendation of another person, state your relationship to that person and how that person is connected to the employer. The person may be a current or past employee, a friend or acquaintance of the person to whom you are writing, or an influential person who is completely independent of your prospective employer. If the person is working for the company, include his or her name, job title, and the department where he or she works. If your letter is a "cold call" letter, written as part of your marketing plan, this paragraph should contain a few well-chosen words about your reasons for being interested in the company.

Second and Third Paragraphs

The second and third paragraphs should contain two or three of your features or strengths that would be of specific interest to the employer. Again, this is an appetizer. Do not try to present your background. Simply state what you are (e.g., a mechanical engineer); what you have accomplished that would be of interest to the person to whom you're writing (e.g., you designed a machine part that was 30 percent more efficient than anything built before); and finally, what you could bring to this employer (e.g., you could develop a similar design for their ongoing project).

Last Paragraph

This is your "close." You reaffirm your interest in the employer and state that what you have heard about the job indicates that your qualifications are such that scheduling an interview is appropriate. You then state that you will call to arrange an interview. Making an assertive statement that you will telephone and then actually doing it will greatly increase your chances of getting an interview. If you sit passively waiting for the employer to call, the chances of being called are slim. Being aggressive and taking the initiative will pay off.

COVER LETTER STRATEGIES

Use your cover letter to focus on those highlights in your resume that will be of special interest to this employer. If your resume is printed and cannot be altered, you can use your cover letter to call attention to the areas that pertain to the job for which you are inquiring. You can customize each cover letter to a particular company without altering your resume.

It is very important to prepare each cover letter individually. Under no circumstances use a preprinted or photocopied cover letter. Consider how you would respond to a letter that began, "Dear Sir or Madam: Your company has been brought to my attention as being a leader in its field … " with the name and address obviously typed in.

Using a word processor, however, you can copy one letter and customize it for each mailing with a minimum of effort. You can use "boilerplate" verbiage and insert information that will give your cover letter the individual focus that will get you an interview.

Your cover letter must be perfect. No misspelled words. No typographical or grammatical errors. The only way to ensure this is to proofread your letter at least twice and have someone else proofread it also. Mistakes stand out and make the statement that you are careless, inattentive, or do not care. Any of these is the kiss of death for a job candidate.

Be sure your cover letter is properly placed. This means it has 1-inch margins at the top, bottom, and sides, plenty of white space, and is balanced on the page. A short letter with all the text on the top and a 6-inch margin at the bottom looks unplanned and un-professional. Make it look good!

SAMPLE COVER LETTERS

The following pages present several sample cover letters. Use these letters as a template, but remember to customize each cover letter to present you in the most effective way to an individual employer.

EUGENE B. ACKERMAN
3456 Lilypad Court
Indianapolis, Indiana 46555
(317) 555-8013

March 30, 1993

Dear Andrew,

I ran into Bob Peters at the IEEE meeting last month. He told me you were doing consulting work for United Tech and suggested I get in touch with you. You may have heard that my company, Dynamics International, has been acquired by a foreign firm and I have decided that it is time to consider other opportunities.

As you know, for the last eight years, I have been designing, fabricating, and installing air ducts and equipment. Since we talked last, I was promoted to General Manager and have become the local expert at developing specs and bidding from architects' drawings.

If you have any suggestions about directions I might take that would assist in my job search, I would be very appreciative. If you can recommend any good search firms that might be of help, please give me a call.

I have enclosed four copies of my resume. Please feel free to distribute them. Since my current company is not yet aware that I am considering a move, I would appreciate your discretion. If you need more resumes or if there is someone you feel I should send one to, please let me know.

Lois and I will be visiting New York next month. I'll give you a call when we get our schedule firmed up. I'll look forward to seeing you then.

Sincerely,

Gene Ackerman

JENNETT ASH
15 Waverly Street
Yonkers, New York 10701
(212) 555-2527

October 24, 1993

Mr. William Johnson
Career Blazers Agency, Inc.
500 Fifth Avenue
New York, NY 10101

Dear Mr. Johnson:

Thank you for discussing current opportunities available to me through Career Blazers. I enjoyed our conversation and appreciate your courtesy in explaining the services available through your organization.

I have considered the position at Children's Hospital that we discussed and would like to find out more about it. Since the resume I left with you does not emphasize my experience as a cytogenic technician, I am enclosing another one that demonstrates my strengths in that area.

I was a Tissue Culture and Cytogenetics Technician with St. Johns Hospital in Yonkers for two years. There I set up and cultured amniotic fluids; prepared, stained, and screened slides; photographed metaphase cells; set up skin biopsies; and prepared cells for growth of amniotic taps. I also participated in diabetics studies related to the development and morphology of the fetal pancreas.

Again, Mr. Johnson, thank you. My experience and education appear to fit the qualifications you described for the position at Children's Hospital. I am looking forward to meeting with the Director of Laboratory Services. If there is any additional information you might need, please call. I'll check back with you around the first of next week.

Sincerely,

Jennett Ash

Enclosure: Revised Resume

JENNETT ASH
15 Waverly Street
Yonkers, New York 10701
(212) 555-2527

October 24, 1993

Mr. William Johnson
Career Blazers Agency, Inc.
500 Fifth Avenue
New York, NY 10101

Dear Mr. Johnson:

I am seeking a position as a Laboratory Technician in a hospital in the metropolitan area. I understand that your organization specializes in the placement of technical medical personnel and may possibly have positions in my field.

I have more than six years experience as a Laboratory Technician and two years as an Emergency Medical Technician where I worked while attending City College of New York. Because I am qualified at several specialties, I have enclosed three versions of my resume that emphasize varying perspectives of my qualifications.

Most recently, I was a Tissue Culture and Cytogenetics Technician with St. Johns Hospital in Yonkers for two years. There I set up and cultured amniotic fluids; prepared, stained, and screened slides; photographed metaphase cells; set up skin biopsies; and prepared cells for growth of amniotic taps. I also participated in diabetics studies related to the development and morphology of the fetal pancreas. I left due to the downsizing of laboratory service functions that resulted from a cutback in funding.

Mr. Johnson, I appreciate whatever you can do for me. My experience and education will qualify me for a number of positions. If there is any additional information you might need, please call. I'll check back with you around the first of next week.

Sincerely,

Jennett Ash

Enclosure: 3 Versions of Resume

62 Rockland Avenue
Boston, MA 22416
July 9, 1993

Mr. George Barfield
Director of Research
Ames & Hunter Inc.
1400 State Street
Boston, MA 21426

Dear Mr. Barfield:

Bob Richards, a colleague of mine at Peabody Chemical, Inc., suggested I send you a copy of my resume.

I am currently employed as a chemical engineer with Lever Brothers, but I am interested in considering other opportunities. At Lever Brothers, I am involved in investigating the effects of catalysts and inhibitors on corrosion and physio-mechanical properties of metals and have designed special apparatus for oxidation testing.

I am looking forward to meeting you so we can discuss how my qualification and background might contribute to the projects now being developed by Ames & Hunter.

I will call you early next week to establish a mutually convenient time for an interview.

Sincerely,

Mary Wilson

Enclosure: Resume

MARGO JEFFERS
180 Palm Court
Boston, MA 62418
(409) 555-0987

October 12, 1993

Ms Patricia Bellow
Bader-Lawrence, Inc.
43 Biscayne Business Park, Ste. 2B
Miami Beach, FL 33180

Dear Ms Bellow:

Elizabeth Rome, of your Boston office, suggested that I send my resume to you. Elizabeth and I worked together at the Hollow Corporation during 1987 and 1988.

There is one item that was not specifically mentioned on my resume that may be of interest to you, my expertise and comprehensive knowledge of Atomic Absorption Spectroscopy (AA), Ion Chromatography (IC), X-ray Photoelectron Spectroscopy (EICA), and various wet chemical methods.

When I was employed by A.B.D., Inc. I was responsible for the design of experimental programs and collection techniques for volatile trace metals, uses and potential applications of ion chromatography, and the effect of ammonia on the sampling and analysis of sulphur oxides and nitrogen oxides.

Much of my experience has been working on projects similar to those that are now on-going at Bader-Lawrence. I am hopeful that you might have opportunities available where I might make a contribution.

I look forward to meeting you. I will call you early next week to set up a meeting. Meanwhile, if you have any questions or need additional information, please call. I can be reached after 6:00 p.m. at (714) 555-4763.

Sincerely,

Margo Jeffers

Enclosure: Resume

11

Job Sources

You now have a first-class resume and are ready to put it to work for you. Your next task is to get your resume into the hands of employers who have immediate needs. There are a variety of methods available to reach such employers. Consider all of them. These methods include:

- Classified advertisements in newspapers and trade journals
- Search organizations
- College and alumni contacts
- Networking
- Personal contacts
- Direct contacts

CLASSIFIED ADVERTISING

The first thing most people who are looking for work reach for is the want-ad section in the local newspapers. They scour those ads daily and eventually add the classified sections of trade journals in their areas of expertise to their sources as well. Why? Because everyone has been exposed to these ads even when they are not in the job market. You cannot pick up a newspaper that does not have a large classified section. Even when you throw away the classified section to get to another section of the paper, you probably see that this is where jobs are advertised. If you were to ask someone where you should look for a job, they would most likely tell you to look in the want-ads.

Are Want-Ads of Value?

The truth is that want-ads are not the best place to find a job. If you conduct a quick survey among friends and acquaintances, you will discover that most found their current positions through other sources. Of course, people do find jobs in the classifieds, but those who do are a small fraction of the total of people hired. Job guidance and career counseling services estimate that 80 percent of all jobs are never advertised. This is what is called the "hidden job market."

If classified ads are of little value in finding a job, why bother to use them at all? Why do employers continue to use them if they are of such limited value? The answer is that ads represent real jobs and hires are generated from them. Employers—like job seekers—labor under the idea that classified advertising is the accepted way to find employees. The reason that you should peruse want-ads is that they are one of many sources you have available and should not be overlooked.

Using Ads to Your Advantage

Although want-ads have the potential to help you locate a job, they offer other benefits as well. Ads contain a wealth of information about the job market, companies, and job skills expectations. They serve as a barometer of the economy, provide general salary information, and give a general idea of the availability of jobs. This information can assist you in adjusting your expectations to the job market reality.

Look under a number of different job titles when reading the classifieds. Employers may list a job under a variety of titles. You benefit from being familiar with all the listings where a job in your area of interest might be found. For example, an opening for an electrical engineer may be listed under "Engineer," "Electrician," "Electronics," "Electronics Engineer," or even "Project Engineers." Likewise, a polymer chemist must check ads for "Scientist," "Chemist," or "Organic Chemist" instead of merely checking under "P" for "Polymer Chemist." An ad may only list a job under a non-descriptive company title, and you may have to read the position description to discover its duties.

Do not be overwhelmed by a laundry list of qualifications that you may not possess. Very few jobs are filled by individuals who meet all the qualifications listed. Ads list the *ideal* qualifications and few jobs would be filled if demands for rigid adherence were followed. Employers hire the people who they perceive can do the job. Address your prospective employer's perception of competence in your resume, your cover letter, and your interview. Qualifications are usually described very specifically to keep the number of respondents and the time required to screen them to a minimum. Employers dream that the perfect person will respond to their ads. In reality, the more attractive the job, the more responses received. Most responses do not fulfill the requirements, and if the job is not filled immediately, employers become more flexible in their demands. Often a job ostensibly requiring 10 years of experience will be filled by a person with only 4 years of experience, with a commensurate drop in the salary offered. But the job will be filled.

In your search of the want-ads, you will find that most ads do not list salaries. Even when they do, often salaries offered are higher or lower than those stated in the ad. Most ads, you will find, demand that you provide your salary history. Remember, the purpose of asking for salary histories is to disqualify candidates. Some ads even state that "resumes received without salary history will not be considered." Although this may sound threatening, an employer who is seriously looking for but having difficulty finding a viable candidate will not discard an inquiry from a reasonably good candidate. For this reason, you may include your salary history if you feel it will work in your favor or omit if you feel it will work against you. If you are a good fit for the position, it is a good bet you will be hearing from the employer regardless of whether you provide salary information. (See Chapter 7, Information Never to Include, for a full discussion.)

Treat Each Ad Individually

Read each ad carefully, determine what the employer is looking for, then answer it. Do not just clip ads, copy the addresses onto an envelope, and send off a stack of envelopes with a resume and cover letter in hopes that one of them will succeed. If possible, tailor your resume to emphasize those areas of your experience that coincide with the requirements listed in the ad. Always include a cover letter. Then, record the name of the company, the position, and the date you sent a resume, either manually or on a computer database. Keep a copy of the ad to track ads that have been repeated—an indication that the job has not been filled and a sign of anxiety by the employer. Follow up your application at a later date. If the ad says "No phone calls," do not be concerned about following up if you have not heard in a week or two. A popular job will receive many responses and your resume may be buried in a stack of resumes. Employers who put statements like this into ads are trying to avoid a flood of calls. They may also have their personnel people doing the screening. If you do so creatively, you may call. For example, call to inquire about some technical aspect of the job. This will give you an opportunity to show your expertise and perhaps open some doors. You can also check out the employer's telephone courtesy.

Blind Ads

Blind or box number ads are those that do not state the name of the employer, but offer only the job description and perhaps some details about the industry. People have been hired from blind ads, but read these also with suspicion.

First, why the secrecy? Why doesn't the employer wish to say who it is? Perhaps the company does not want current employees to know it is looking for new employees or does not want its competitors to know the type of talent it is seeking. It may also be, however, that the company is not looking at all but is just testing the waters to see what kind of talent is available. A blind ad may also be a company's way of seeing how many of its own employees respond to an attractive opportunity. A blind ad also may have been placed by a search organization building its resume file with a juicy sounding position as bait.

Finally, a blind ad enables employers to discriminate without revealing themselves. If you are rejected, you may never know why or by whom. People do get hired from blind ads, but the unknowns may make it less likely. Caution is advised in putting a lot of time into answering blind ads.

The Timing of Ads

Although newspapers run ads in every issue, certain days of the week feature special expanded want-ad sections. For example, the *New York Times* lists employment opportunities and services on Wednesdays and Sundays, the *Los Angeles Times* publishes its largest classified section on Sundays, while *The Wall Street Journal* publishes the bulk of its want-ads every Tuesday. As you become more involved in your job search, you will come to know where to look in which papers or journals for the types of jobs you are seeking. You will also become adroit at reading and interpreting ads. Then, just about the time you begin to get really good, you will find a job!

A Final Word on Ads

As stated, ads are not the primary way people are hired for most positions. However, people do get jobs through ads. Do not be discouraged if jobs for which you feel imminently qualified get no response or if it takes weeks or even months before you hear. Employers often take their time hiring. They are not likely to be as rushed as you or feel the urgency that you feel to begin the hiring process. It is not unusual for job hunters to receive responses from ads they have answered several months after they found jobs.

Keep in mind that job hunting is a cumulative effort. Use several sources, stick to your search process, and your hard work will pay off. Remember, everyone who looks for work will find work.

SEARCH ORGANIZATIONS

About 9 percent of jobs are found through some form of search or employment organization. Search organizations, in addition to the possibility of being directly responsible for finding you a job, can be useful in providing you with job hunting skills training and tips. They can also help keep you abreast of the latest information in your sector of the job marketplace.

Search organizations vary in their emphasis and methods of doing business. Here are just a few with which you should be familiar:

- Employment agencies
- Contingency search and placement
- Retained search (executive recruiters)
- Outplacement services
- Contractors
- State agencies
- College placement offices

Employment Agencies

Employment agencies are generally licensed by the state in which they operate. This depends on the laws of the different states and is usually to protect applicants who pay fees for being placed in positions. Most technical and professional level positions are fee paid, that is, any fees or commissions are paid by the hiring company. Employment agencies' sole source of income is through the fees they earn for successfully finding the right person for a job opening.

Employment agencies normally handle jobs from those slightly above entry level, two years plus experience to mid-range technical and professional positions. Some agencies offer middle management and first-or second-line supervisory positions, but few high-level and executive management positions.

Employment agency counselors think in terms of the responsibilities and salaries of the positions they are representing. It is easiest for them to fill a position with someone who already has had the level of responsibility specified by the employer. If you want to increase your job responsibility or salary, you must be sure to have a job consultant who is willing to work with you to achieve that goal and capable of doing so. Before committing your time to any employment agency, have a serious conversation about your goals with the person who will represent you in a job search. If he or she is not willing to work with you to get you the job you desire, look elsewhere. Keep in mind, however, that your expectations must be realistic. Do not expect employment professionals to work miracles on your behalf. Consider your real capabilities and potentials—they are what your job consultant has to work with.

Select a search or placement professional who understands the technical field you are in, has a good track record in placing people in your field, and understands the details of the technical placement marketplace. If the consultant has worked in your specialty, that is a plus.

Contingency Search and Placement

Contingency search and placement organizations typically are used by employers to fill middle-to upper-level positions in specific industries or job skill areas. The term *contingency* means they receive a fee for their services only if they are successful in filling their employer client's requirements. *Search* means they are given a specific assignment by an employer and then they search the marketplace for suitable candidates. This is similar to the activities of a retained search organization. Placement means they will represent qualified candidates to employers who may be able to use their talents.

Contingency and retained search organizations operate in similar markets. However, the contingency recruiter tends to be more aggressive because there is no payment until the search is successful. For maximum results when dealing with a contingency recruiter, seek one who is willing to work at placement. Also contact search recruiters who are specialists in your technical field. Having your name in their files will make you available as opportunities arise.

Retained Search (Executive Recruiters)

These recruiters search for high-level (and high-salary) professionals and are paid through a retainer or prepaid fee by their employer clients. Normally you do not find them, they find you. If you contact them directly, they will ask you to send a resume that will be filed. If they receive an assignment and you meet the qualifications, you may be called. However, unless they have a current assignment that fits you, they will not be very responsive to your needs. Do not contact retained search organizations unless they work exclusively in your field.

Outplacement Services

Outplacement is a service provided to companies who are laying off, downsizing, reorganizing, or terminating personnel. Outplacement organizations provide training in job search skills such as interviewing, networking, and salary negotiation. They also assist in resume preparation and mailings. Outplacement eases the corporate conscience by offering laid-off employees help in getting relocated, and provides a useful service to be used to the maximum. Remember, use every resource in your job search.

Contractors

Contracting involves taking a temporary assignment on contract instead of a permanent job. When a large number of technical people are out of work or looking for new opportunities, contracting offers an alternative to permanent jobs that may be scarce or unavailable. You get to work in your specialty and will likely get a much higher rate of pay for your services. The work, however, may be uncertain. There may be no benefits such as health insurance or vacations, and there is no career path. Also, you may be expected to travel and be away from home for extended periods of time.

Do not reject contracting as a possibility. Many employers, rethinking the way their workforces are organized, are reluctant to hire permanent staff for technical projects and may have no place for permanent employees when the project is complete. They find bringing on contract professionals to perform the work a better solution. Even though the hourly rate paid to contractors is higher, using temporary help is more economical than hiring permanent people, paying them benefits, and being faced with the possibility of letting them go at the completion of a project.

State Agencies

The United States Employment Service (USES) is a network of more than 2,000 offices set up to serve both the job seeker and the employer seeking personnel. Because the agency is operated by the government, there are no fees to either the employer or the job seeker.

USES works much the same as a private employment agency performing recruiting and placement services. However, most corporate executives contacted indicated they would be more likely to use the services of a private organization because they felt they offered a more effective and professional service. The strength of USES is not in its placement services but in its specialized services such as counseling, testing, and outreach.

State agencies also offer help to professional and technical job seekers. The stereotype of the state employment office is one of indigents lined up to collect their unemployment checks and the only jobs being offered menial, minimum wage positions. However, many states offer substantial services to higher-level job seekers. For example, California, through its Employment Development Department, sponsors Professional Resource Groups that specifically address the employment concerns of managerial and technical people. They meet weekly, network, train participants in job search skills, and provide moral support. In many ways they are equal to professional outplacement services. And, they do not charge job hunters for their services.

College Placement Offices

Many entry-level engineering, scientific, and computer information systems positions are filled through college placement offices. If you are a recent graduate or about to graduate, avail yourself of this valuable service. Even if you did not graduate from or attend a particular college, check their job listings and counseling services.

Another college and university resource that can be extremely useful is the alumni association, particularly as a networking source. The wise job hunter, however, uses every contact to its maximum.

MAXIMIZE YOUR USE OF SEARCH ORGANIZATIONS

1. Try to establish rapport with the counselor or consultant with whom you will work. People who like each other work better together; people who do not like each other do not work well together. Work to develop a positive relationship.

2. Be systematic about finding the right sources to help you. Ask friends whether they know an agency or search professional they might recommend. Check the Yellow Pages for search and placement organizations that specialize in your field. Check with professional associations such as the National Association of Personnel Consultants for recommendations.

3. Confide in your consultant. Follow his or her advice. Ask questions if you do not know why you are being asked to perform some task. Build the relationship by telling your consultant about possible job leads or candidates who might fit other job assignments.

4. Do not bother your consultant with "How's it going" or "What have you done for me lately" calls. If your consultant is part of a private firm, he or she is financially motivated to succeed on your behalf. If the consultant hears something, you will hear, too.

5. Provide your consultant with plenty of resumes in the format and form requested. Do not expect the consultant to rewrite or reformat your resume.

COLLEGE AND ALUMNI CONTACTS

As mentioned previously, a valuable source of entry-level positions is the placement office of your college or of colleges in the area where you would like to work. For the experienced person or the new graduate, your college alumni association can be very helpful. You can locate alumni by company and position and feel comfortable contacting them directly. If they cannot help you directly, they may be in a position to refer you to someone who can. Do not worry about appearing too aggressive. You are looking for a job. Everybody has been in that position at some time or another. Do not hesitate to remind them of that fact if they act as if they are going to cut you off and not help you. Remember, your objective is to get a job.

After you have made a contact, offer to drop your resume off at the contact's office or even take him or her to lunch. Let the person know you would appreciate advice and direction. Everyone likes to be sought after. A prominent or well-placed alumnus is a powerful ally.

NETWORKING

Networking is the oldest and remains the best of all job hunting methods. Simply stated, when you network you ask people you know whether anyone they know has work available. This method of building a set of contacts is the one most likely to lead you to your ideal job. Take an informal survey. You will find that the majority of those you talk with found their current jobs because someone they knew steered them to the person or employer who hired them.

Networking builds on itself. Eventually you begin to realize that the number of contacts who can help you is endless. Do not overlook any of them. Begin by telling everyone you meet that you are hunting for a job, and ask whether they know someone who can direct you to places or people who are hiring. If your current job situation makes such public knowledge risky to your position, be more judicious in your networking.

Networking is the most popular source of jobs because it is a personal, face-to-face experience. The person who recommends you to a company is a credible reference and serves as an immediate reference check. You would be more comfortable hiring someone who had been referred to you by a trusted and respected acquaintance, and so would most employers.

Start your networking by telling your family, friends, and social acquaintances. Call former employers, current and former fellow workers, and colleagues. Do not be embarrassed to ask for leads. Your friends and colleagues have been in similar positions and understand that any kind of help is welcome.

Become a joiner. Join professional societies where you can meet other professionals in your field. There is no secret to networking. It is the process of extending your ability to make contacts by working through other people.

Of course, telling everyone you are looking for a new opportunity is a shotgun approach to your job search. A more effective search is more focused. If you know a company or organization you might like to work for, ask your contacts whether they know someone who works for that company. Then call that person, using your contact's name as an introduction, and ask for the name of the person who manages the department where you would like to work.

The networking method is a very effective way of finding a job. It gives you a "voice at court" who can speak for you and provide a reference on your behalf. Employers will feel more comfortable with candidates who are referred by someone they know. This is no guarantee that you will get the job, but you will be in a better position.

12

The Interview

As mentioned throughout the text, the goal of preparing and effectively using a high-quality resume is to get a face-to-face interview. The interview is the single most important event of any job search. All the tasks you have done—preparing and mailing resumes, networking, answering ads, or researching companies—are preliminary steps leading to the interview.

Now that your resume has done its job and you have been scheduled for an interview, do not screw it up!

The resume you have prepared is a tool that can be of great help during and after an interview. Prepare your resume carefully and use it skillfully. The purpose of this book is to guide you as you construct a resume (or resumes) that will give you an introduction to prospective employers that results in interviews. To achieve this, you must continually review and fine tune your resume. You must view the resume as a dynamic document and continually seek ways to improve upon it.

HOW YOUR RESUME HELPS YOU IN YOUR INTERVIEW

Possibly the worst thing you could do on a job interview is fall short of the expectations the employer has fixed in his or her mind from the information on your resume. Your resume has given your prospective employer a mental picture of you. The expectation is positive and will work to your advantage.

When employers call candidates for interviews, they want them to be the right choices for the jobs they offer. The last thing an employer wants is to waste time talking to unqualified people. If the claims you made on your resume are not supported in the interview, you will be out the door. Your goal in an interview is to represent in person the competence and the qualifications the employer saw in your resume. The key word here is *competence* because you must project to the employer that you can do the job. No employer will hire you if he or she harbors even a hint of doubt that you can do the job.

PREPARING FOR THE INTERVIEW

Having researched employers before sending them your resume, you will know something about the company you are going to interview with. That initial research, however, is superficial compared to what you must do before your scheduled interview.

Gather Information About the Company

Learn as much as you can about the prospective employer's business. Know the name of the person with whom you are interviewing and the names of other executives as well. Become knowledgeable about their product line or the services they offer.

1. Visit a library and review business directories there to see what they say about the company. Directories such as *Dun & Bradstreet, Standard & Poor's*, or *Moody's Industrial Manual* contain financial information, company histories, and the names of officers and executives. Check the business sections of your local newspapers as well as the trade journals that cover specific industries. Find out about upcoming mergers, acquisitions, new products, or problems. If you discover something negative, don't write the company off. Mentally file the information and keep your eyes and ears open. At the appropriate time during your interview you can raise questions about issues that concern you.

2. Talk to your bank or stockbroker about the companies who invite you in for an interview. They may be able to get you a copy of an annual report or they may have current information that is not yet published.

3. Drop by the company a few days before the interview. You may be able to obtain an annual report or product literature from the receptionist in the visitor's lobby. You can also observe the general demeanor of the company. Is it friendly? Does it look organized and orderly? Do the employees look happy? If the company has a cafeteria, visit it if possible.

You can pick up a wealth of information there. You will hear gripes and praises, and have an opportunity to see the company in an unguarded way. Most companies do not open their cafeterias to nonemployees, but if you have the opportunity, take it.

4. It is likely that you will be scheduled by someone other than the person who will interview you. You will probably be called by personnel or the hiring manager's secretary or assistant. If so, tactfully ask the person scheduling your appointment about the position for which you are being considered. Ask about the duties of the job. Ask who you will be reporting to and whether the position includes any supervisory activities. Learn as much as possible, including any negative aspects of the job such as excessive travel or long hours. Approach such negative issues with caution, however. During your initial contacts with a prospective employer, try to be upbeat and positive. Asking about the job duties shows that you are more interested in the job than in pleasing yourself. Asking about negative issues too early in the interview process may turn off employers. If something of a negative nature comes up, make a mental note to discuss it to your satisfaction when you receive a job offer.

Review Information About Yourself

1. Review your resume prior to your interview. During the interview a prospective employer will refer to your resume frequently, using it as a guide for the interview. Be prepared to discuss and expand upon each point in your resume.

2. Be prepared to fill out an employment application. When the interview is scheduled, request that an employment application be sent to you, so you can fill it out without feeling rushed. You can choose your words more carefully and fill in all the blanks without having to call back with missing information after the interviw. If you are not able to get an application before the interview, come prepared to provide your social security number; names, addresses, and telephone numbers of references; the exact dates of your previous jobs; job titles and duties; names of supervisors; and your educational history.

The information on your employment application form should agree with that on your resume. Any discrepancies cause doubts, raising unnecessary questions. Fill out the application as thoroughly as you can. Always fill in the employment history, and never write "see resume." Be neat and thorough. If you are careless and sloppy in filling out an employment application, the employer might assume that your work would follow the same pattern or that you are not enthusiastic about the job.

3. Try to anticipate the questions you will be asked during the interview. If there are issues that might be of concern such as being fired from your previous job or why you want to leave your current job, you must be prepared to respond to any inquiry in a confident and assured manner. It is likely that you may not have to answer any of the questions for which you prepare, but it is vital that you prepare just in case.

Make a Good Impression

The impression you make when you meet a potential employer for the first time will be the one on which you will be judged as suitable for a job. This initial meeting may last only a few minutes, but it is of critical importance in the hiring process.

You want to literally put your best foot forward. Your dress, posture, attitude, vocabulary, and body language must coalesce to give the interviewer a favorable first impression.

Dress for the Interview

Dress appropriately. Do not overdress or wear expensive clothing. Dress conservatively. For a man, this means a dark blue or gray suit with a white or light blue or gray shirt with a tie that is color coordinated and in good taste. Women should follow the same rule. A blue or dark gray skirt or business suit with a white blouse is acceptable dress in any business situation. A woman does not have to dress as a man to present a proper business appearance, but she must present a businesslike image. Anything less than conservative business clothing for an interview will not get the job done.

Here are some do's and don'ts for dressing for an interview:

1. Do dress conservatively in good-quality clothing. A sure-fire model for what is considered appropriate for any company can be found in the company's annual report. The picture of the CEO or president will show you a person whose dress is always acceptable.

2. Do always shine your shoes and wear freshly pressed clothing. A sloppy appearance says that your work is probably sloppy also.

3. Don't wear casual clothing to an interview. This means no blue jeans, open-collared shirts, sweaters, or sports jackets. Even if the dress code at the company is casual, wear a business suit to the initial interview. After you have been hired, you can wear what everyone else wears.

4. Don't wear extreme styles. Remember, the key word in dressing for an interview is *conservative*. An interview is not the place to make a fashion statement.

Have the Proper Interview Attitude

The only acceptable attitude in a job interview is enthusiastic. Someone in the company saw enough in your resume to invite you in for an interview. That person hopes you will be the right person for the job. Technical professionals are in short supply, and a top professional will always be in demand regardless of the state of the job market. Approach the interview with the feeling that the company needs you, and be optimistic that you will get the job.

Enthusiasm in the interview means showing an interest in the company, the position, and the interviewer. If you have done your homework in researching the company and the person you are interviewing, you will be able to convey interest and enthusiasm.

You must also convey sincerity. This means be yourself. Be honest and open. Do not try to act like someone you are not. This will work in your favor. A primary reason for a face-to-face interview is to determine your personal "chemistry," that is, what kind of person you are and whether or not you will fit into the company. It is not likely you could keep up a pretense for an entire interview. Even if you could, you could not do so on the job. You want to be hired for who *you* are.

Be Prepared to Handle Nervousness

Even the most experienced and confident technical professionals become apprehensive at the thought of a job interview. There is good reason for this. A job interview is not a common occurrence. It is stressful. There is a great deal riding on its outcome: the way you spend your working life, your income, and a number of other things.

Most interviewers have been in the position of being interviewed, understand the feelings you are experiencing, and will make allowances for nervousness. An experienced interviewer will very likely do his or her best to put you at ease.

The most effective way of overcoming nervousness is to be prepared for your interview. The more you know about the company and the person you are interviewing, the higher your level of confidence and the less reason you will have to be nervous.

If you have not visited the company, it is also a good idea to make a dry run to the company location a couple of days before your scheduled interview. You will learn about where to park, where to enter the building, how long it will take you to drive to the appointment, and a number of other logistical considerations. The last thing you want to do is to run late, have difficulty finding a place to park your car, and report for the interview sweating and out of breath. Consider the impression that a wheezing, sweaty candidate will make on a prospective employer.

On your dry run, make some observations about the company. Do you see happy faces or do you hear complaints? Is there a feeling of organization or chaos?

The Day of the Interview

To make your interview a success, plan ahead.

- Lay out your clothes the night before, making sure they are clean and pressed.
- Give yourself plenty of time to dress and get ready before you leave for the appointment.
- Unless the interview is taking place during breakfast, lunch, or dinner, eat before you go. Never go to an interview on an empty stomach. Also, do not eat in your interview clothes. The last thing you need is jelly down the front of your shirt.
- Hang your coat in your car to keep it free of wrinkles. Do not wear it as you drive.
- Plan your arrival for about 15 minutes before your interview time. If you run into traffic or a problem, you will have a small buffer. If not, you can use those 15 minutes to visit a restroom and make a last-minute check on your appearance.

THE INTERVIEW

Prior to your interview, engage in a mental exercise that will boost your confidence. Say to yourself, "I am the best person for the job and I will convince the employer of that fact. I *will* be offered this job!" Thoughts may cross your mind such as "What if this isn't the job for me?" or "Shouldn't I give the job more thought before making such a commitment?" The answers are "Of course you won't take the job if it's not right" and "Certainly you should give any job a great deal of thought."

By thinking confidently, you will believe you are the best person for the job and radiate that belief. An employer will not offer you a job without being convinced that you can do it. By radiating confidence, you infect the employer with your conviction. This may not be the job you will ultimately take, but preparing yourself to act positively and confidently in an interview will ensure that you will make the right moves when the right job comes along.

Opening Moves

The first person you are likely to meet at the employer's place of business will be a receptionist or secretary. The positive impression you make on the employer starts with this person. Be on your best behavior with everyone you meet in the company.

When you meet the employer for the first time, step forward and greet him or her in a positive manner. Listen to how the employer greets you. If the employer greets you casually, you would be proper in responding casually. If the employer first welcomes you without stating his or her first name, respond more formally. If he or she responds by asking you to use a first name, take your lead from that.

Psychologists state that the first minutes of a meeting between two people are the most important in establishing a bond or relationship. Initial impressions carry far more impact than impressions made later in the relationship. It is vital that you make a positive impression right away. When you shake hands, make your grip firm but not crushing. In U.S. society, a limp handshake is viewed as a weakness in both men and women.

Take a seat when your host offers one, but you may sit in any case when the interviewer sits down. Lean forward a bit in your chair to indicate interest and attention. Leaning back or slouching is often interpreted as a sign of disinterest. Body language is an important part of the communication process. Use it to your advantage.

The Interview Communication Process

If there were only one piece of advice that could be passed along about communication, it would be the importance of listening. In an interview you will be expected to talk more than the interviewer.

However, your listening skills must be sharp and alert. Often the interviewer will monopolize the conversation. Go along with it and respond to the interviewer by asking positive, motivational questions about the job, the company, and how you might contribute. Avoid asking questions about vacations, sick days, or medical plans. If the interviewer brings up these or related subjects, listen to the information but always return to the positive issues.

Handling Tough Questions

The most difficult questions you will be asked during your interview are those requiring more than a simple yes or no. These open-ended questions can be answered in a variety of ways. They provide a wealth of information about you and open the interview to more probing. Although there is no way of knowing what an interviewer might ask, certain questions invariably pop up regularly during interviews.

The following list includes a few of those questions. Study them carefully and consider the variations that might be directed at you. Keep in mind that the interviewer will be listening not only for the content of your response but also for how you answer, that is, the sincerity you project as well as your ability to think on your feet. Have someone read these questions to you while you respond extemporaneously.

- What are your short- and long-range goals?
- Why do you want this job?
- What are your qualifications for this job?
- What did you like most or least about your last job?
- What can you tell me about yourself?
- What motivates you?
- How do you motivate others?
- How do you relate to your peers, your superiors, and your subordinates?
- How do feel about working overtime or on weekends?
- What are your major strengths and weaknesses?
- How well do you work under pressure?
- Can you give examples of problems you have solved and how?
- What did you learn while working at your last position?
- How did you get along with your previous boss and staff?

- How do you feel about working for a member of the opposite sex?
- How do you feel about working for a younger person?
- Do you prefer to work alone or as a part of a team?
- What magazines, newspapers, and books do you read?
- How do you spend your free time?
- Why should we hire you instead of someone else?
- Where do you expect to be in your career 5 or 10 years from now?
- How have your education and past experience prepared you for this job?
- Describe three personal or professional accomplishments that have given you the greatest satisfaction.
- What have you learned from your mistakes?
- What do you know about this company?

These questions can be answered in a variety of ways. To help focus on a question, you may wish to ask the interviewer to clarify a point. For example, if you were asked "What can you tell me about yourself?" you might narrow the scope by asking the interviewer whether there was some specific area he or she would like to hear about.

Questions About Money

Avoid the subject of money until you are certain that you will receive an offer. When the subject of money is discussed during an interview, it tends to divert the discussion from more important job issues. The interview is part of the screening process; you do not want to be screened out until all the facts have been considered.

Never bring up the subject of money during the interview. If the employer brings it up, turn the conversation back to the job and its requirements. The employer usually asks a question such as "What are your salary expectations?" Reply that you are more interested in the opportunity and you are confident that the starting salary the employer offers will be fair and reasonable. This statement implies you will receive an offer and that the offer will be subject to review.

Employers often ask for a salary history, and it is appropriate for you to provide it. The standard question usually asks your current salary. The primary purpose of this question is to determine whether what you are currently earning is above the range that the employer expects to pay for the new position. Actually, your current salary has nothing to do with the value of a new job to you. When addressing a potential position in terms of the salary refer to what the employer is paying employees who are performing similar tasks and have backgrounds and capabilities similar to yours.

THE JOB OFFER

If you are offered a job, consider other factors in addition to salary. The new position might offer benefits such as health care, vacation, and the opportunity for advancement or challenge. Consider job stability, the work environment, the people you will be working for and with, and the location. Although money is important, a high salary cannot make you happy in your work.

Most employers will not extend a job offer during the first interview. If you should receive an offer during the interview, however, thank the interviewer and say that you will give it careful consideration, that you would appreciate a couple of days to think it over. Do this even if your first impulse is to hug and kiss the employer and accept the offer. Changing jobs and accepting a new one is an emotionally charged process. Always take time to reflect and let the dust settle.

More likely, you will receive a job offer a day or so after your interview. When you receive this offer, thank the employer, express your appreciation for the confidence expressed in you, and say you will think it over. You are not playing a game of hard to get with the employer. By thinking it over you will have an opportunity to evaluate the offer logically. If you have been unemployed for a long time, the thought of working again may be exciting and you may leap at a job offer you will later regret. The amount of time you wait may be a day or even a few hours. Never take more than a week, because the employer's selection process stops until you make a decision. The employer wants you, and you owe that employer an answer within a reasonable time.

You can delay acceptance of an offer in order to renegotiate the terms of the offer. Although the offer may be somewhat flexible, the dollar amount usually has been derived from compensation tables and job descriptions and so is relatively firm. A more fruitful approach is to negotiate to have the job reclassified. The result is the same: You get more money and a promotion before you start the job. If you are being represented by a search or placement organization, the consultants would conduct these negotiations, acting as a neutral buffer between you and the employer. Also, because their fee is usually based on the salary you will receive, they are motivated to get you the highest salary offer possible.

When you do accept the offer, do it verbally and in writing. Call the person who extended you the offer to express your appreciation and say that you look forward to becoming a part of the organization. If the personnel department, not the person for whom you will be working, extended the offer, you will want to call your new boss and express your pleasure about the opportunity. Follow this call with a written acceptance that includes a restatement of the offer and its terms, including the start date of your new job.

FINAL DO'S AND DON'TS FOR INTERVIEWING

1. Do be at the interview on time. In fact, get there 15 minutes early. Promptness is an absolute must. If you are delayed for any reason, telephone immediately to reschedule the interview.

2. Don't schedule more than two interviews a day. When interviews are scarce, you may be tempted to schedule as many as you can get. One in the morning and one in the afternoon are plenty.

3. Do fill out the application form completely.

4. Don't smoke even if the interviewer offers you a cigarette. Most workplaces are moving toward smoke-free environments and nonsmokers are more welcome than smokers.

5. Do be polite at all times. Never lose your cool. If the interviewer becomes antagonistic, stay calm; he or she is may be trying to evaluate how you respond to stress.

6. Don't interview the interviewer. Dominating the interview may boost your ego, but it won't get you the job.

7. Do ask questions. If there are any aspects of the job, its responsibilities, or company benefits that are unclear, ask for the information you want.

8. Don't be vague. Think about the questions before you answer. Make your answers clear, definite, and concise. Remember, if you do not know an answer, "I don't know" is an acceptable response.

9. Do be professional. Maintain good eye contact and use positive body language.

10. Do ask for the job. When you leave the interview, the employer should have no doubt as to your interest in the job.

11. Do send the interviewer a note thanking him or her for the time the interview took and reiterating your interest in the position and the company. This should be done within one day following your interview.

13

Keeping Records

Good recordkeeping during your job search is important. Keep a record of each resume sent and the dates of your calls and interviews. Record the results of each call and interview, and keep track of follow-up activities. Keep copies of all letters you sent, networking calls you made, follow-ups, and results. Do not trust a record of all your job search activities to your memory. By maintaining a written record of your activities you add order and discipline to an important effort that otherwise can become buried in an avalanche of activities. The information in your records can also help you reach better decisions throughout your job search.

MANUAL RECORDKEEPING

The easiest way of maintaining a record of your direct mail campaign, answers to classified ads, resumes you have mailed, and your networking correspondence is to make a copy of each letter and resume you send out, noting the dates of each and the results of follow-up phone calls and interviews. Keep these in a file folder along with a calendar noting the scheduled dates and times of interviews. A calendar helps you avoid scheduling two interviews at the same time.

A second manual system is set up on an 8 1 / 2- by 11-inch sheet of paper with column headings across the top. A suggested format is shown in Figure 13-1.

The third manual system uses 4- by 6-inch index cards. Although the format is different, the information is the same as in the other manual systems. Figure 13-2 shows a sample layout for the

Resume Mailing	Follow-Up Phone Call	Interview	Thank-You Letter	Job Offer	Confirmation or "No Thank You, But" Letter
Name	Date	Date	Date	Yes	Date
	Results	Time		No	Letter Type
Company		Interviewer			
Address		Results			
Date Sent					

FIGURE 13.1   Job Search Tracking System Using 8 1 / 2- by 11-Inch Sheets of Paper.

JOB SEARCH ACTIVITY TRACKING SYSTEM

RESUME MAILED TO: **MAILED:**
Mr. Richard Rowe   9/23/93
Director of Engineering
Systems, Inc.
424 Park Place
Buford, PA 21370

PHONE CALL: _____
                            (indicate date)
(Note Results) _____

INTERVIEW: _____
                    (indicate date, time, and interviewer)
(Note Results) _____

THANK-YOU LETTER: _____

JOB OFFER: _____

CONFIRMATION OR "NO THANK YOU, BUT" LETTER: _____
                                    (indicate date and letter type)

FIGURE 13.2   Record of Resume Contact and Interview Activity Using 4- by 6-Inch Card.

card. This system works well for large mailings. Have the template printed on the index cards by a "quick print" service rather than typing each one yourself.

Keeping a record of your job campaign activities will enable you to keep track of the companies you have contacted and ensure that you do not send a resume to a company twice or schedule interviews at conflicting times. It is also an excellent way to review the progress of your job campaign at any point in time.

AUTOMATED RECORDKEEPING

If you have a personal computer, you have the option of employing technology to assist you in your job search and recordkeeping. Computers can dramatically increase the power and efficiency of any job search.

Using the versatile software available for personal computers contributes to effective tracking of resumes, calls, interviews, and networking activities. The word processing capabilities described in Chapter 8 can bring a new dimension to your resume.

Recordkeeping Software

The following software is useful for your recordkeeping:

- Data base management systems (DBMSs)
- Personal information managers (PIMs)
- Contact management systems (CMSs)

Select software on the basis of what works best for your particular need.

Data Base Management Systems (DBMSs)

DBMS software comprises application programs that control the organization, storage, and retrieval of data in a database. A DBMS permits the user to design custom records, organize the information in those records, and exhibit it in a custom designed format. The main advantage of a DBMS is the flexibility to design a system that fits your needs. Purchase the software from an established company to ensure a reasonable level of support and updates to new software versions as they become available. The main drawback to DBMS programs is that they often require some knowledge of database design or programming to use them to their fullest capability. If you are already using one of the popular DBMS packages, you probably have the ability to design a job search database application to fit your needs. If you are not currently a user, choose a software better fitted to your needs.

Popular DBMS packages include *dBASE IV*, *Paradox 4.0*, and *Paradox for Windows* from Borland and *Foxbase* and *Access* from Microsoft, Inc. Why do the same software manufacturers produce competing packages? Because the more powerful companies buy up the competition. For example, Borland purchased Ashton-Tate and their popular DBMS, *dBASE*, the industry standard for more than 10 years and the one with the largest user base. Their competing product, *Paradox*, is viewed by most computer industry observers as a more advanced and powerful DBMS. A similar situation existed with Microsoft. Their newest product is called *Access*, the newest big name entrant to the DBMS software marketplace. However, they also purchased *Foxbase*, a popular DBMS designed along the dBASE model. The suggested retail price of these systems is approximately $700 each. The price you will pay at a discount software store or through mail order is about 25 percent less.

Personal Information Managers (PIMs)

A PIM is a DBMS that is structured with pre-designed file formats and a collection of utilities and tools designed to help you administer your business and personal life. A PIM allows you to enter names, addresses, phone numbers, and appointments, and to keep notes on meetings, projects, and activities. A PIM will also maintain a "to-do" list and structure reports on the data in the system. The advantage of a PIM to you as a job hunter is that it is ready to go as soon as you install the package. The file records come already set up. You can redesign the format simply. Generally, PIMs are easy to learn and operate and are quite suitable for organizing a job search. A strength particular to PIMs is their ability to display topics, names, and companies, by searching on a word, phrase, or subject.

Representative software packages include *Pack Rat* from Polaris and *GrandView* from Symatec. The suggested retail price is around $400 but, like the DBMS, dramatically less from discounters.

Contact Management Systems

Contact Management Systems (CMSs) are similar to PIMs except their focus is on prospecting and sales, and they contain a low-level word processing system that permits you to format form letters and customized letters. The system automatically records calls, appointments, and letters into history files. Some contact managers interface with word processing programs to interactively create and modify custom resumes. A contact manager is also a DBMS that is designed for a specialized set of tasks.

ACT! for Windows is a CMS used to maintain personal, contact, and business files. This packaged software is very well suited to recordkeeping in a job search. *ACT! for Windows* allows you to keep and update files on prospective companies, technical managers, or network contacts. Each record has 70 customizable fields that can be coupled with pop-up screens for frequently used entries. An example of the "contact record" screen is shown in Figure 13-3. Date-stamped notes, activities, and a comprehensive history log are also attached. *ACT! for Windows* also permits you to schedule calls and appointments, and keep a comprehensive history log. An alarm can also be set to remind you of calls or appointments. You can log appointments with a single keystroke (Figure 13-4), and then display them on a daily log (Figure 13-5) or monthly calendar (Figure 13-6).

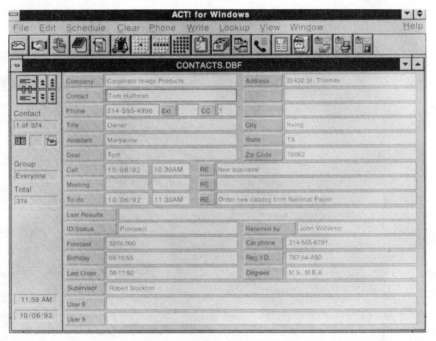

FIGURE 13.3 Contact Record Screen—*ACT! for Windows.* Each contact record of *ACT! for Windows* includes two screens that can be customized to fit specific needs. Notes, activities, and a history log are attached to each contact.

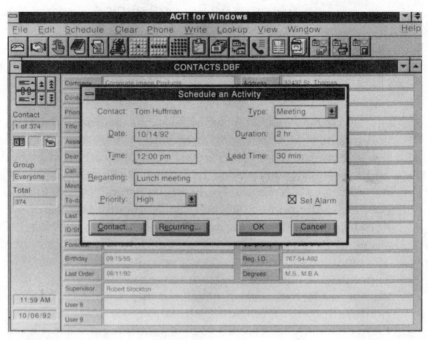

FIGURE 13.4 *ACT! for Windows* Activity Schedule Pop-Up Window. *ACT! for Windows* permits unlimited scheduling of activities and reminder alarms.

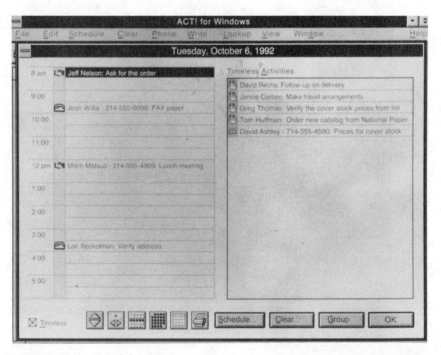

FIGURE 13.5 *ACT! for Windows* Activity Schedule Daily Planner. *ACT! for Windows* provides daily, weekly, and monthy views of scheduled activities.

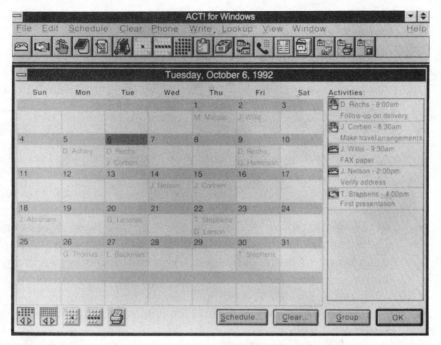

FIGURE 13.6 *Act! for Windows* Activity Schedule Monthly View. *ACT! for Windows* offers a monthly overview of all scheduled activities.

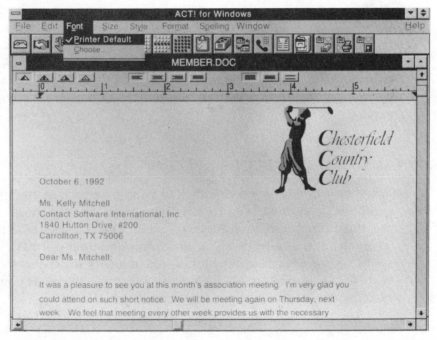

FIGURE 13.7 *ACT! for Windows* Correspondence Screen. *ACT! for Windows* built-in word processor offers WYSIWYG, spell checking, and mail merging.

ACT! for Windows' scheduling utility features the ability to prioritize activities, utilize automatic conflict scheduling to avoid more than one activity in a time period, and enter user-definable lead times and activity durations. A list of appointments, tasks, and "to-do's" is available with a keystroke.

The word processing and letter writing feature is also a handy feature for job hunters. Within *ACT! for Windows* is a limited but sufficient word processor that offers a spell checker and "What You See Is What You Get" (WYSIWYG) capability. This means that type fonts and formatting are shown as they will be printed. *ACT!* also offers quick memo, form letter, envelope, and fax cover sheet generation. Dates, names, addresses, salutations, and closings are automatically inserted in the correspondence. Graphics from other applications can be pasted in for letterheads and special effects. A sample of *ACT! for Windows* generated correspondence is shown in Figure 13-7.

Report generation in *ACT! for Windows* includes reproduction of address and phone lists, activity reports, contact and history reports, day/week/month calendar printing, and easy custom report creation.

ACT! for Windows is also available in a DOS version, for non-Windows and the Macintosh. A networked version is available for the PC versions.

ACT! for Windows is not the only CMS available, but it is currently the most popular of that type software and it is well supported by its manufacturer, Contact Software International. The manufacturer's suggested retail price is $495, but like other packages, it can be purchased through discount channels at a substantial savings.

A FINAL WORD ON RECORDKEEPING

Whether you choose to keep track of your job search manually or by computer, you must keep records. The more detailed those records are, the more information they will yield. You can use these records to analyze the effectiveness of certain resume formats or responsiveness of employers. In addition, many of the costs associated with a job search are deductible from your income tax and the records you keep will help substantiate your claims.

The personal computer is no longer a mere novelty or even a choice as a business tool; it is a necessity. A professional in any field who cannot use rudimentary tools such as word processing, spreadsheets, or data base systems is severely handicapped considering the competition now in the job marketplace. When you prepare your resumes and letters by personal computer, you are making a statement of your computer proficiency. If you use a typewriter and your resumes and letters contain the typos, errors, and formats common in the typewriter era, you date yourself and make a negative statement.

14

Sample Resumes

An effective way to develop a high-impact resume is to study resumes that have worked in the past. This chapter contains a list of 10 resume types and formats that will be useful as you select formats that will work best for your particular situation.

Also included in this chapter are a collection of sample resumes and examples of various resumes representing different jobs and positions. These sample resumes may give you ideas about formats, layouts, and forms to help you prepare your resume. Do not copy the examples of job descriptions or work experience, even if those descriptions are very close to your experience. Your resume must be an original work written and designed by you. Never have someone else write it for you. The want-ads are full of resume writing services that promise resumes that get jobs. These services can provide typing and printing services; do not allow them to write your job description or do any of the things that only you can do best.

Search organizations receive dozens of resumes daily and see the best and the worst. The formats used by resume writing services reveal who created the resume to the experienced resume reader. Although their appearance is satisfactory, the description they provide for you usually leaves much to be desired. For example, they seldom talk about your technical skills and the equipment you have worked with in specific technical terms. You may be described in glowing generalities with an abundance of "power" and "action" words thrown in, but while the words sound good, the reader is left wondering what it is you do. There is nothing wrong in using strong, descriptive language to describe what you do. However, it must be in the proper context. Only you know what you do well enough to create an accurate resume. Resumes that are obviously written by someone else, particularly by a professional resume writing service, are generally not given much weight by prospective employers, while a thoughtfully prepared resume that demonstrates technical knowledge is usually well received.

Develop your worksheet, choose a format, and start writing. Use this first effort as a basis to rewrite, reformat, and rewrite again until you have a product that pleases you. Proofread your resume until you never want to see it again. If you are using a word processor, use the spell checker and grammar checker. When your resume looks good to you, show it to someone else who will give you the harshest constructive criticism. If possible, choose professionals in your technical field who have hired people themselves.

After your resume has been criticized, scrutinized, and edited, repeat the review process. Do this until you have a resume that you feel absolutely certain will get you a face-to-face interview with prospective employers.

Several of the sample resumes are shown in two formats: One is a single font, single type size, typewriter format; the other is an enhanced version produced on a computer using different types of fonts with features such as boldface, italic, and shading. Judge for yourself which best suits your needs.

These sample resumes are real resumes. Only the names and contact information have been altered. Where real company names are used, no endorsement of those particular organizations is meant, nor is there any implication that those companies might be hiring. These resumes have performed as they were designed to do—that is, to get the person they represent in front of an employer for an interview.

In review, here are some basic do's and don'ts of successful resume writing:

- Do keep your resume brief, one or at the most, two pages. If you must go over two pages, make sure the first page grabs the reader's attention.

- Do use 8 1/2- by 11-inch paper. This is standard business correspondence size. No odd sizes or foldover brochure style formats. Stick to a pale neutral color (white, cream, gray, or beige) and no loud or dark colors. Use only one side of the paper. No fancy binders. No company letterhead.

- Do place your name, address, and telephone number in a prominent and conspicuous position. Do not include your current business number unless your employer knows that you are looking or you want him to know.

- Do keep your objective statement brief and to the point if you decide to use one. An objective statement should be relevant to your experience, education, and the job for which you are sending your resume. A better tactic is to include your job and career objectives in your cover letter rather than your resume.

- Do list your current or most recent job first, and then continue in reverse chronological order.

- Do account for all dates of employment. Leave no unexplained gaps.

- Do give your resume more punch and power by using action verbs.

- Do list your education in reverse chronological order. List the most advanced degree first, continuing to the bachelor's or associate's degree.

- Do include the dates all degrees were received.

- Do make your resume visually inviting. Use wide margins and plenty of white space. If you have a word processor or desktop publishing capability, consider enhancing your resume as shown in the various examples. Do not overdo enhancements. You want to make your resume more appealing, not show how many typefaces you have on your computer.

- Do, if your resume is more than one page, staple the pages in the upper left hand corner. Put your name on each page.

- Do proofread your resume. Look for misspelling, grammatical errors, and typos. If you prepare your resume on a PC, use the spell checker and grammar checker, then read it over after it is printed. Have someone else, preferably a technical professional, proofread your work.

- Do read your resume from the employer's point of view. Remember, a resume that you like may not necessarily be a successful resume. Only the resume that the employer likes and finds interesting enough to call you in for an interview can be called successful.

- Don't include a photograph, salary information, reasons for leaving previous jobs, or names and addresses of references. Don't include personal information.

- Don't include long, hard-to-read paragraphs. Proofread your resume for readability as well as format and content.

A few final caveats before you embark on preparing your resume. Your best resume is within you. However, a good resume is not all you need to get a job; it is only one very important tool you must use and use well to succeed. Remember, a job search is the hardest job you will ever undertake—and the most important.

RESUME TYPES AND FORMATS

Some resume books do not discuss resume types because they present only one type as the "winning" resume. It must be said again and again that there is no single "winning" or "best" resume. The winning resume is the one that gets you interviews.

Resume Types

Chronological

This is the most common resume type and generally the type most used by job seekers and preferred by employers. It lists work history by company or organization with reference to the periods of employment. The chronology is normally in reverse order, listing the most recent experience first.

Functional

The functional resume emphasizes specific areas of competence. It is useful when you have held a number of jobs or there are gaps in your work chronology. For example, if you were emphasizing an area such as design, you would describe your skills and accomplishments without reference to when or for what organization they were performed.

Narrative

The narrative resume appears to be more of a biographical description than a resume. It is written in a continuous prose or narrative style. This format is often used by companies to represent their employees to clients. For personal use, refer to yourself in the first person.

Focused

The focused resume is similar to a functional or narrative resume but focuses on a particular job, career path, or company. It is normally written for a specific job in a known company.

Organizational

This resume lists your jobs by company emphasis. It may or may not contain your work chronology. This format is often used with a chronology when you have worked for a number of well-known companies but have moved frequently between jobs.

Creative

Creative resumes are unique because they are the creation of the job seeker. They may vary in layout, type, use of graphics, size, color, and content. In the business world, they are usually considered taboo, but some circumstances may demand that a more creative approach be used. A touch of creativity can help in any resume. Generally, however, the best approach is to remain conservative and conventional.

Enhanced

This is a hybrid format used in many of the sample resumes in this book. It is combined with either the chronological or functional format. Enhancement involves using your word processor to employ larger or unique type styles with names, headings, job titles, and other areas you wish to stand out. It makes use of italics, boldface, and underlining for specific emphasis. The key to success is a tasteful application of the special features you have available in your word processor.

Biographical Summary

This is much like the narrative format except that it uses short, incomplete sentences to highlight the job hunter's background. This format is often used by third parties representing the job hunter (agencies, headhunters, etc.). It can be very effective with employers with whom you have had previous contact or who know you.

Mini-Resume

The brief statement that you might include in a chronological or functional resume is actually a "mini-resume" that can be used as an insert or enclosure in inquiry correspondence.

Bullet/Accomplishment

This is an outlining style using "bullets" or "dingbats" such as ❑ ❖ ▼ ◆ ● ■ ○ ☞ or any other symbol to emphasize a point or accomplishment. This approach is a hybrid that can be used in combination with any other format type.

A Word About Resume Types and Formats

Despite the various formats available, the resume types that have been best received are the chronological or functional formats with an enhanced layout. Employers are accustomed to seeing these types of resumes. Enhancing a resume with fonts and effects to make it more appealing and emphatic makes a solid, well-received resume style even better.

SAMPLE RESUMES AND EXAMPLES

Following are sample resumes of technical professionals in a variety of format examples. All the resumes are actual resumes and only the name of the person represented and the contact information have been changed.

Section 1 contains Sample Resumes—Plain and Fancy. These resumes demonstrate the formatting and visual effects possible with popular word processors and tools. Section 2 is a collection of resumes that actually got interviews for the people they represented. Section 3 shows two special types of resumes that can be used to your advantage when networking.

Space considerations have dictated that not all technical professions and jobs be represented. The sample resumes were chosen because they represented positive examples and techniques in resume development. Keep these tips in mind as you prepare your resume.

- Use the sample resumes as models and examples. Remember, what might work for someone else might not work for you.

- Make your resume your creation. Copying someone else's resume, regardless of how good it might be, will not make it yours. You are a unique person and you want to express your differences positively.

- The guidelines and advice in the book concerning resume writing have been developed as a result of experience reviewing thousands of resumes. Most resumes in the marketplace are poor examples of the professionals they are supposed to represent. Don't let yours be one of them.

- Do not be afraid to experiment and be creative. Before you settle on a type of style and format, try some different fonts and formats. You are the only one who will see what you do until you determine that you have created your best.

- Get several second opinions on your resume. Ask for a truthful review. Discover the flaws in your resume before rather than after you mail it.

- Do not settle on only one resume. Develop several with variations on each one. Customize each one for specific employers.

How to Use the Sample Resumes

The resumes that follow represent a variety of formats and technical occupations. Most are labeled at the top of the page with the profession represented, and in some cases the type of resume is shown. For example:

Electrical Engineer (Enhanced)

Examine these resumes to see which delivers the message you want to give to employers. That is, what is most effective in saying you are qualified for the position offered. What format and style compels the reader to call you in for an interview?

Sample Resumes— Plain and Fancy

The resumes in this section represent the many varieties that can be created using the various fonts and effects available on the major word processing software packages. The resumes are of people who have used them successfully in their job searches. In several instances there are multiple resumes for the same person that demonstrate the effects of enhancing a plain resume to make it more appealing.

The first resume is that of "Stella McGowan." Stella is a fictitious name, as are the other names in this section. However, Stella's resume is real. A "Commentary and Critique ..." precedes Stella's two resumes. Use these to evaluate the other resumes reproduced here.

Be critical *and* creative. Look at the samples, get ideas, then put those ideas into action as you prepare your resume.

COMMENTARY AND CRITIQUE OF STELLA McGOWAN'S RESUME

The following two resumes are those of "Stella McGowan."

The first resume appeared in the first edition of *Best Resumes for Scientists and Engineers*. It was created with a typewriter using pica 12-point type. Although it is satisfactory and presents an adequate appearance, there is nothing about it that stands out.

The second resume was created on a word processor using a variety of effects to create a more striking and visually appealing resume. Stella McGowan's name was enlarged to 22-point Garamond bold. Her contact information was set in True Type Ariel bold to make it stand out. The section headings were enlarged and made Burlesque bold. Ms McGowan's coursework in the education section has been italicized for emphasis. Job titles in each experience description are bold.

The enhanced resume also fits on one page, while the original required two. This was accomplished by using the reformatting capabilities of the word processor.

You are the final judge as to whether your resume conveys your message. It took less than five minutes to perform the enhancements shown here. Always enter your resume into a word processor so you may enhance, modify, or reformat.

Several of the resumes in this sample section are presented in single typeface style as though on a typewriter. An enhanced version is also presented so you may see the creative application of a word processor and make a comparison. If you are in doubt as to the effectiveness of this resume creation technique, remember you will be competing with many other resumes. Give yourself every break available!

Stella McGowan
180 Lode Street
Stanford, Connecticut 71429
(615) 812-6368

EDUCATION:

9/82 - Present   <u>BA in Human Biology</u>, Bridgeport University, Darien, Connecticut. Coursework included computer science, economics, anthropology, and statistics.

7/85 - 8/85   Boston College, Boston, Massachusetts. Summer school coursework in Spanish.

EXPERIENCE:

1/86 - 6/86   <u>Research Assistant</u>, Health Promotion Unit, Department of Gynecology and Obstetrics, Boston University Medical Center. Planned and implemented animal and epidemiological studies concerning women's health care.

5/85 - 6/86   <u>Investigator</u>, Mid Peninsula Fair Housing Council, Palo Alto, California. Investigated complaints of discrimination in housing by conducting telephone and personal interviews.

10/84 - 6/86   <u>Office Assistant</u>, Division of Otolaryngology, Boston University Medical Center. Provided clerical and administrative support for division personnel.

10/83 - 6/84   <u>Lab Assistant</u>, Plasmid Reference Center, Stanford University Medical Center. Prepared nutrient medium for bacteria cultures and maintained lab supplies.

OTHER EXPERIENCE:

Secretary, typist, cashier, sales clerk.

ADDITIONAL INFORMATION:

Able to read, write, and speak Spanish.

Enjoy playing jazz and classical music on the piano and
flute.

Member, Stanford Committee of Black Performing Arts.

REFERENCES

Available upon request.

Stella McGowan

180 Lode Street
Stanford, Connecticut 71429
(615) 832-6568

EDUCATION:

1986	**BA in Human Biology**, Bridgeport University Darian, Connecticut. Coursework included *Computer Science, Economics, Anthropology,* and *Statistics.*
1985	Boston College, Boston, Massachusetts. Summer School Coursework in Spanish.

EXPERIENCE:

1/86 - 6/86	**Research Assistant:** Health Promotion Unit, Department of Gynecology and Obstetric, Boston University Medical Center. Planned and implemented animal and epidemiological studies concerning women's health care.
5/85 - 6/85	**Investigator:** Mid Peninsula Fair Housing Council, Palo Alto, California. Investigated complaints of discrimination in housing by conducting telephone and personal interviews.
10/84 - 6/86	**Office Assistant:** Division of Otolaryngology, Boston University Medical Center. Provided clerical and administrative support for division personnel.
10/83 - 6/84	**Lab Assistant:** Plasmid Reference Center, Stanford University Medical Center. Prepared nutrient medium for bacteria cultures and maintained lab supplies.

OTHER EXPERIENCE: Secretary, Typist, Cashier, Sales Clerk

ADDITIONAL INFORMATION:

Able to read, write, and speak Spanish.
Enjoy playing jazz and classical music on the piano and flute.
Member, Stanford Committee of Black Performing Arts.

REFERENCES: Available upon request.

Sylvia White
18 Kittrige Street
Los Angeles, CA
(213) 555-8840
(213) 555-8000

CAREER OBJECTIVE

Seeking a challenging position in the Aerospace Industry or related Electronic field where experience and formal academic training will be fully utilized. Position must offer opportunities for professional development.

EDUCATION

San Francisco State University   San Francisco, CA
Bachelor of Science Degree, Electrical Engineering   June 1991

Advanced Engineering Courses:

Materials of Science Engineering	Strength of Materials
Electronic Communication (1&2)	Statics
Digital Electronic Design	Systems Analysis
Microcomputer Programming	Circuit Analysis
Engineering Analysis	Thermodynamics
Electric Machines	Control Systems (1&2)
Analog Electronic Design	Computer Logic Design
Heat Transfer	Fortran IV (77)

WORK

Curtis Research & Testing Corporation   Los Angeles, CA
Test Engineering   September 1990 - Present

Scope of Activities:
Provide quality control testing, qualification programs, research studies and design support projects such as fatigue investigations, joint strength, lap shear, and load allowable testing.

Specialties:
Mechanical Fasteners: Comprehensive testing and evaluation programs on complete range of fastening devices in accordance with applicable military, federal, ASTM, SAE, Prime and major industry specs and standards. Test procedures observing requirements of Mil-Std-1312, ASTM F606, Mil-N-25027, and Fed Test Method Std. No. 151.

Plain Bearings: Perform evaluation of TFE lined and plain spherical bearing to military standards, including oscillating radial load testing. Fatigue testing of rod and bearing.

Fatigue: Routine and research fatigue evaluations of threaded and non-threaded fasteners to Mil-Std-1312 and NAS1069. Fatigue Life Studies include S-N curve development on advanced materials, weldments, and structural joints. Special programs at elevated temperature.

Static: Static strength testing with capabilities of 400,000 pounds. Tensile strength and double/single shear testing. Full range of extensometers for yield strength studies. Elevated temperature testing with capabilities of 1800 degrees F, and rapid heat rate testing up to 100 degrees F per second. Special cryogenic (sub-zero) testing.

Stress Rupture: Stress rupture testing with capabilities of 1800 degrees F for military, AMS, and industry spec requirements for evaluation of high performance materials and components.

Vibration: Special test support for evaluation of mechanical fasteners under vibration test outlined by government and industry standards.

Environmental: Salt spray, humidity, and torque studies related to fasteners and mechanical components.

Metallurgical Laboratory: Micro and macro examination of fasteners. Areas of investigation include fillet radius, lap seams, grain flow, grain size, carburization, decarburization, and crest laps.

Torque-Tension: Developing torque-tension design data on current and advanced fastening systems for industry and government.

REFERENCES

Professional and personal references available upon request.

Sylvia White
18 Kittrige Street
Los Angeles, CA
(213) 555-8840 (H)   -   (213) 555-8000 (O)

EDUCATION

San Francisco State University   San Francisco, CA
Bachelor of Science Degree, Electrical Engineering   June 1991

Advanced Engineering Courses

Materials of Science Engineering   Strength of Materials
Electronic Communication (1&2)   Statics
Digital Electronic Design   Systems Analysis
Microcomputer Programming   Circuit Analysis
Engineering Analysis   Thermodynamics
Electric Machines   Control Systems (1&2)
Analog Electronic Design   Computer Logic Design
Heat Transfer   Fortran IV (77)

WORK

Curtis Research & Testing Corporation   **Los Angeles, CA**
Test Engineering   **September 1990 - Present**

Scope of Activities

Provide quality control testing, qualification programs, research studies and design support projects such as fatigue investigations, joint strength, lap shear, and load allowable testing.

Specialties

Mechanical Fasteners: Comprehensive testing and evaluation programs on complete range of fastening devices in accordance with applicable military, federal, ASTM, SAE, Prime and major industry specs and standards. Test procedures observing requirements of Mil-Std-1312, ASTM F606, Mil-N-25027, and Fed Test Method Std. No. 151.

Plain Bearings: Perform evaluation of TFE lined and plain spherical bearing to military standards, including oscillating radial load testing. Fatigue testing of rod and bearing.

Fatigue: Routine and research fatigue evaluations of threaded and non-threaded fasteners to Mil-Std-1312 and NAS1069. Fatigue Life Studies include S-N curve development on advanced materials, weldments, and structural joints. Special programs at elevated temperature.

Static: Static strength testing with capabilities of 400,000 pounds. tensile strength and double/single shear testing. Full range of extensometers for yield strength studies. Elevated temperature testing with capabilities of 1800 degrees F, and rapid heat rate testing up to 100 degrees F per second. Special cryogenic (sub-zero) testing.

Stress Rupture: Stress rupture testing with capabilities of 1800 degrees F for military, AMS, and industry spec requirements for evaluation of high performance materials and components.

Vibration: Special test support for evaluation of mechanical fasteners under vibration test outlined by government and industry standards.

Environmental: Salt spray, humidity, and torque studies related to fasteners and mechanical components.

Metallurgical Laboratory: Micro and macro examination of fasteners. Areas of investigation include fillet radius, lap seams, grain flow, grain size, carburization, decarburization, and crest laps.

Torque-Tension: Developing torque-tension design data on current and advanced fastening systems for industry and government.

REFERENCES

Professional and personal references available upon request.

Lawrence Moore
105 Riverview Drive
Dallas, Texas 75213

(214) 826-7156 (Home)
(214) 535-5000 (Work)

POSITION DESIRED

Applications engineer for factory automation and
robotics systems.

SUMMARY

Effective educational background in systems
engineering, with a specialization in human factors.
Four years diverse experience in general systems and
human factors engineering. Extensive experience in
integrating new subsystems into existing manufacturing
systems to improve overall system performance.

EXPERIENCE

3-91 to Present   Ergon Logikos
                               Bayfield, Colorado

Proprietary applied R&D on scientific software

1-89 to 2-91   Union Carbide, Inc.
                               Ft. Worth, Texas

Human factors Engineering (H.F.E.) advisor to Vice
President of Operations for monitoring results of first year
activities in improving assembler engineering disciplines
within Operations Department. Project efforts included
classification of H.F.E. as cross parametric analysis of
classifications to thirty percent salary increase during
first eighteen months with company.

1-86 to 9-88   Baylor University Laboratories, Inc.
                           Austin, Texas

Head Lab Technician responsible for developing methods
and procedures for vision experiments. Also was
involved with collection and statistical analysis of
experimental data. Coordinated three other lab
technicians efforts. Conducted informal lectures and
supervised undergraduate students during Perceptual
Methods course lab sessions. Guest lecturer for
advanced perception classes.

EDUCATION

8-88   Wright State University
                           Dayton, Ohio

B.S. in Systems Engineering, major in Human Factors

Industrial courses taken in Value Engineering, Quality
Circle Training, and Logistics Technology and
Management

Member of Human Factors Society and International
Ergonomics Society

REFERENCES

Upon request

Lawrence Moore

105 Riverview Drive
Dallas, Texas 75213

(214) 826-7156 (Home)
(214) 535-5000 (Work)

POSITION DESIRED

Applications engineer for factory automation and robotics systems.

SUMMARY

Effective educational background in systems engineering, with a specialization in human factors. Four years diverse experience in general systems and human factors engineering. Extensive experience in integrating new subsystems into existing manufacturing systems to improve overall system performance.

EXPERIENCE

3-91 to Present   **Ergon Logikos**
                                       Bayfield, Colorado

Proprietary applied R&D on scientific software

1-89 to 2-91   **Union Carbide, Inc.**
                                       Ft. Worth, Texas

Human Factors Engineering (H.F.E.) advisor to Vice President of Operations for monitoring results of first year activities in improving assembler engineering disciplines within Operations Department. Project efforts included classification of H.F.E. as cross parametric analysis of classifications. Received a thirty percent salary increase during first eighteen months with company.

1-86 to 9-88   **Baylor University Laboratories**
                                       Austin, Texas

Head Lab Technician responsible for developing methods and procedures for vision experiments. Also involved with collection and statistical analysis of experimental data. Coordinated three other lab technicians' efforts. Conducted informal lectures and supervised undergraduate students during Perceptual Methods course lab sessions. Guest lecturer for advanced perception classes.

EDUCATION

8-88   **Wright State University**
                                           Dayton, Ohio

BS in Systems Engineering, major in Human Factors

Industrial courses taken in Value Engineering, Quality Circle Training, and Logistics Technology and Management

Member of Human Factors Society and International Ergonomics Society

REFERENCES

Upon request

Ralph Gardner
438 Pineforest Avenue, Apt. 35
Jackson, Mississippi 39421

(601) 555-2832 (Home)
(601) 555-3742 (Work)

EDUCATION   Ph.D., Polymer Chemistry,
                   University of Southern Mississippi, 1992
                   Hattiesburg, Mississippi

                   MS, Chemistry, Loyola University, 1983
                   New Orleans, Louisiana

                   BS, Chemistry Loyola University, 1981
                   New Orleans, Louisiana

WORK EXPERIENCE

The University of Southern Mississippi   JUNE 1989 to MAY 1992
Department of Polymer
Science, Hattiesburg, Mississippi

Postdoctoral Research Associate: Conducted research on the photodegradation
of Polyimides, Polyetherketones, and Polyarylates. Developed and standardized
a technique for studying the Laser Ablation of Polyimides. Published 2 research
papers in scientific journals. Presented 5 research papers in international and
national conferences.

Faxton Finishes   JULY 1986 to JUNE 1989
Memphis, Tennessee

Project Chemical Engineer: Developed an alkyd resin for use in automotive
refinishes. Developed equivalents for 4 paint additives, which were either
monopoly supplies or imported ones. Learned the methods of synthesis of
different types of polymers that are used as binders in paints. Directed the
activities of the Advanced Instrumentation Laboratory, which involved analytical
method development and extensive polymer characterization using IR, NMR, UV,
GC, HPLC, and GPC. Supervised 1 senior chemist and 4 junior chemists (all MS
degree holders). Directed the standardization of computerized color matching
of paints.

Midcontinents Research Institute   JULY 1983 to JUNE 1986
St. Louis, Missouri

Project Assistant: Performed research on the modeling of Ammonium
Perchlorate combustion. Developed a computer program to enable calculation
of parameters that were required for the modeling studies.

PROFESSIONAL PUBLICATIONS
AND PRESENTATIONS

List attached

Ralph Gardner

438 Pineforest Avenue, Apt. 35
Jackson, Mississippi 39421

(601) 555-2832 (Home)
(601) 555-3742 (Work)

EDUCATION   Ph.D., Polymer Chemistry
University of Southern Mississippi, 1992
Hattiesburg, Mississippi

MS, Chemistry, Loyola University, 1983
New Orleans, Louisiana

BS, Chemistry Loyola University, 1981
New Orleans, Louisiana

WORK EXPERIENCE

The University of Southern Mississippi   **JUNE 1989 to MAY 1992**
Department of Polymer
Science, Hattiesburg, Mississippi'

**Postdoctoral Research Associate:** Conducted research on the photodegradation of Polyimides, Polyetherketones, and Polyarylates. Developed and standardized a technique for studying the Laser Ablation of Polyimides. Published 2 research papers in scientific journals. Presented 5 research papers in international and national conferences.

Faxton Finishes   **JULY 1986 to JUNE 1989**
Memphis, Tennessee

**Project Chemical Engineer:** Developed an alkyd resin for use in automotive refinishes. Developed equivalents for 4 paint additives, which were either monopoly supplies or imported ones. Learned the methods of synthesis of different types of polymers that are used as binders in paints. Directed the activities of the Advanced Instrumentation Laboratory, which involved analytical method development and extensive polymer characterization using IR, NMR, UV, GC, HPLC, and GPC. Supervised 1 senior chemist and 4 junior chemists (all MS degree holders). Directed the standardization of computerized color matching of paints.

Midcontinents Research Institute
St. Louis, Missouri

Project Assistant: Performed research on the modeling of Ammonium Perchlorate combustion. Developed a computer program to enable calculation of parameters that were required for the modeling studies.

PROFESSIONAL PUBLICATIONS
AND PRESENTATIONS

List attached

Computer Operator

Luther Farnington
29888 Society Drive
Norwalk, CA 91700
Home: (714) 555-9299

PROFESSIONAL OBJECTIVE

Seeking a responsible position as Computer Operator.

SUMMARY OF QUALIFICATIONS

Extensive education and experience on the following systems: IBM, Digital, Control Data, and Computer Vision.

PROFESSIONAL EXPERIENCE

COMPUTER OPERATOR   ROCKWELL INT'L   DOWNEY, CA   4/88–PRESENT

Responsible for the efficient operation of all computers and peripherals at the CAD/CAM site. Includes: Digital VAX systems 11/785, 782, and 730; Control Data Cyber 170 series 835; Computer Vision 11/16 system and Calcomp and Versatec plotters. Second shift operator working unsupervised. Duties include: Daily and weekly backups of all systems, daily and weekly accounting, executing plots, archiving as needed, restoring user files, authorizing users, monitoring the systems, and troubleshooting problems as they arise.

LAB TECHNICIAN   GOULD MSD   OXNARD, CA   8/84–3/88

Responsible for quality control and manufacturing of resistors, using a photo-electric process, and various stages of electronic assembly.

EDUCATIONAL DATA

COMPUTER LEARNING CENTER, ANAHEIM, CA

Course work in IBM operating systems OS and DOS including command languages, job control language, system utilities, RPG programming and operations management. Grade Point Average 4.0. Graduated 11-8-87.

SAWYER BUSINESS COLLEGE, POMONA, CA

Course work in business math, typing, vocabulary, business machines, and PBX. Grade Point Average 4.0. Graduated 9-82.

LUTHER FARNINGTON

29888 Society Drive, Norwalk, CA 91700   Home: (714) 555-2922

PROFESSIONAL OBJECTIVE

Seeking a responsible position as Computer Operator.

SUMMARY OF QUALIFICATIONS

Extensive education and experience on the following systems: IBM, Digital, Control Data and Computer Vision.

PROFESSIONAL EXPERIENCE

COMPUTER OPERATOR   ROCKWELL INT'L, DOWNEY, CA   4/88 - PRESENT

Responsible for the efficient operation of all computers and peripherals at the CAD/CAM site. Includes: Digital VAX systems 11/785, 782, 780, and 730; Control Data Cyber 170 series 835; Computer Vision 11/16 system and Calcomp and Versatec plotters. Second shift operator working unsupervised. Duties include: Daily and weekly backups of all systems, daily and weekly accounting, executing plots, archiving as needed, restoring user files, authorizing users, monitoring the systems, and troubleshooting problems as they arise.

LAB TECHNICIAN   GOULD MSD, OXNARD, CA   8/84 - 3/88

Responsible for quality control and manufacturing of resistors, using a photo-electric process, and various stages of electronic assembly.

EDUCATIONAL DATA

COMPUTER LEARNING CENTER, ANAHEIM, CA

Course work in IBM operating systems OS and DOS including command languages, job control language, system utilities, RPG programming and operations management. Grade Point Average 4.0. Graduated 11-8-87.

SAWYER BUSINESS COLLEGE, POMONA, CA

Course work in business math, typing, vocabulary, business machines and PBX. Grade Point Average 4.0. Graduated 9-82

Computer Programmer/Data Administrator

JOHN EASTMAN TOLLING
258937 Lawson Avenue
La Mirada, CA 90638

(213) 555-7707

EDUCATION: BS, Computer Information Systems,
California State Polytechnic University, Pomona, 1992

EXPERIENCE: ROCKWELL INTERNATIONAL CORPORATION
Garden Grove, CA

Jun 87 to
Present

PROGRAMMER/ANALYST: Worked in the areas of Data Base Administration and Data Security. Utilized the DB2 DBMS with COBOL to write reports regarding the security and types of data access granted each user. Maintained the ACF2 Dataset and Transaction Rules for all corporate systems. Participated in the design of Application Security for General Employment Management Systems. Also used the IBM PC and Compact Deskpro. Utilized dBASE III, RBase 5000, and LOTUS 1-2-3.

The mainframe systems environment was IBM 3081/MVS/XA with TSO/SPF, JES3, ACF2 Security, and the DB2 Data Base Management System. Programming was performed in COBOL.
(Apr 1991 to Present)

ROCKWELL INTERNATIONAL CORPORATION
Seal Beach, CA

PROGRAMMER/ANALYST: Designed COBOL for Data Administration that kept track of Systems Loggings and Violations. Designed Structure Charts and Generic Programs. Wrote several COBOL programs that performed standard functions such as Sorts, Updates, Report Writers, Binary Searches, and File Merges. Programs resulted in a savings of two programming hours per day following implementation.

The systems environment was identical to Rockwell Garden Grove Facility, i.e., IBM/MVS/XA with DB2, COBOL, and supporting software utilities.
(Jun 1987 to Mar 1991)

JOHN EASTMAN TOLLING
258937 Lawson Avenue
La Mirada, CA 90638

(213) 555-7707

EDUCATION: BS, Computer Information Systems,
California State Polytechnic University, Pomona, 1992

EXPERIENCE: ROCKWELL INTERNATIONAL CORPORATION
Garden Grove, CA

Jun 87 to
Present

PROGRAMMER/ANALYST: Worked in the areas of Data Base Admini-stration and Data Security. Utilized the DB2 DBMS with COBOL to write reports regarding the security and types of data access granted each user. Maintained the ACF2 Dataset and Transaction Rules for all corporate systems. Participated in the design of Application Security for General Employment Management Systems. Also used the IBM PC and Compact Deskpro. Utilized dBASE III, RBase 5000, and LOTUS 1-2-3.

The mainframe systems environment was IBM 3081/MVS/XA with TSO/SPF, JES3, ACF2 Security, and the DB2 Data Base Management System. Programming was performed in COBOL.
(Apr 1991 to Present)

ROCKWELL INTERNATIONAL CORPORATION
Seal Beach, CA

PROGRAMMER/ANALYST: Designed COBOL systems for Data Admininstration to track Systems Loggings and Violations. Designed Structure Charts and Generic Programs. Wrote COBOL programs that performed standard functions such as Sorts, Updates, Report Writers, Binary Searches, and File Merges. Programs resulted in a 2 hour savings of programming hours per day following implementation.

The systems environment was identical to Rockwell Garden Grove Facility, i.e., IBM/MVS/XA with DB2, COBOL, and supporting software utilities.
(Jun 1987 to Mar 1991)

GEORGE NASH
1236 Williams Avenue
Glendale, CA 91256
(213) 555-9834 * (213) 555-4000

EXPERIENCE:

1/91 - Present   Field/Construction Engineer
American Import, Inc.
San Diego, CA

Direct construction projects, verify compliance with blueprints
and structural design, supervise safety practices, and purchase
job site materials.

9/86 - 1/91   Teaching and Research Assistant
University of California
Berkeley, CA

Assisted in instruction of undergraduate and graduate classes in
absence of professor. Graded papers and tutored undergraduate
and graduate students. Participated in research efforts dealing
with noise attenuation of jet engines.

3/89 - 11/90   Design Engineer
Levy & Levy, Inc.
North Hollywood, CA

HVAC design and energy calculations for residential and
commercial buildings, negotiations with vendors and suppliers,
coordination with customers and city officials.

6/86 - 2/89   Manager
AT&T, Inc.
La Mirada, CA

In charge of employee work schedules and evaluations,
business operations including cash flow, banking and financial
reports, repair/remodeling of equipment, and customer relations.

EDUCATION   Master of Science in Mechanical Engineering, Jan. 1991
                University of California, Berkeley, CA

                Bachelor of Science in Mechanical Engineering, Sept. 1988
                Hollywood State University, CA

                E.I.T. - State of California

                M.S. Thesis: Kinematic Analysis of the Planetary Gear Trains

                Computer Languages: BASIC, FORTRAN, PASCAL, C

REFERENCES   Available on request

GEORGE NASH
1236 Williams Avenue
Glendale, CA 91256
(213) 555-9834 * (213) 555-4000

EXPERIENCE:

1/91 - Present   **Field/Construction Engineer**
American Import, Inc.
San Diego, CA

Direct construction projects, verify compliance with blueprints
and structural design, supervise safety practices, and purchase
job site materials.

9/86 - 1/91   **Teaching and Research Assistant**
University of California
Berkeley, CA

Assisted in instruction of undergraduate and graduate classes in
absence of professor. Graded papers and tutored undergraduate
and graduate students. Participated in research efforts dealing
with noise attenuation of jet engines.

3/89 - 11/90   **Design Engineer**
Levy & Levy, Inc.
North Hollywood, CA

HVAC design and energy calculations for residential and
commercial buildings, negotiations with vendors and suppliers,
coordination with customers and city officials.

6/86 - 2/89   **Manager**
AT&T, Inc.
La Mirada, CA

In charge of employee work schedules and evaluations,
business operations including cash flow, banking and financial
reports, repair/remodeling of equipment, and customer relations.

EDUCATION   **Master of Science in Mechanical Engineering**, Jan. 1991
University of California, Berkeley, CA

Bachelor of Science in Mechanical Engineering, Sept. 1988
Hollywood State University, CA

E.I.T. - State of California

M.S. Thesis: Kinematic Analysis of the Planetary Gear Trains

Computer Languages: BASIC, FORTRAN, PASCAL, C

REFERENCES   Available on request

Electrical Engineer

Gary Stern
185 Salem St.
Boston, Mass. 61246
(617) 798-4683
(617) 845-2800

CAREER OBJECTIVE

Electrical engineering position involving application
of Digital Electronics, Electronic Communication,
Control Systems, Microprocessor, Power Systems Analog
Circuit, Electromechanics, AC & DC Motors,
Alternator, and Generator.

EDUCATION

1980 to	B. S. degree from Lehigh University
1983	Electrical Engineering major and Electronics minor.
1974 to	A. S. degree from University of Electronic
1976	Communication. Electronic Communication major.

SKILLS

Analog Circuit and Signal Analysis, Digital and
Control Systems, Electromechanics, Power Systems
Energy, Microprocess, Programmable Controller.
Ability acquired by design project; both academic and
personal experience.

EXPERIENCE

1985–present	Trichman Engineering, Inc. Control Design Engineer. Design software with logic program (ladder diagram logic relay) and hardware interface with programmable controller. Bailey network 90, Modicon 884 and Micro 84, Control AC motor (Speed switch), reactor temperature (thermostat), damper (open and close), blower and conveyor system.

1984 to   U.S. Rubber, Inc.
1985   Control Design Engineer. Design software with logic
            program (ladder diagram logic relay), interface with
            design hardware for several applications (Analog
            input, Analog output). Programmable controller
            (Allen-Bradley system) for industrial (PLC 2/20,
            2/15, 2/30, PLC 3). Motion control, torque control
            for DC/AC motors, temperature and pressure control
            used for tool machines.

1983 to   A. & Z. Torque Systems, Inc.
1984   Product engineering. Interfaced to manufacturing and
            customer, analog and digital systems involved,
            design, redesign, modification, DC motors, control
            system, amplifier, power supply, encoder, decoder and
            pulse modulator.

1974 to   American Airport
1978   System Engineer. Tested equipment of radio receiver
            transmission, Electronic Communication, Analog and
            Digital System, Power System, Electromechanics, AC &
            DC motors, Dynamics, Alternator, and Generator.

REFERENCES

Available on request.

Gary Stearn
185 Salem Street
Boston, Massachusetts 61246
(617) 555-4683
(617) 555-2800

CAREER OBJECTIVE

An electrical engineering position involving application of Digital Electronics, Electronic Communication, Control Systems, Microprocessor, Power Systems Analog Circuit, Electromechanics, AC & DC Motors, Alternator, and Generator.

EDUCATION

BS, Electrical Engineering (Major) and Electronics (Minor), 1983
Lehigh University

AS, Electronic Communications, 1976
University of Electronic Communication

SKILLS

Analog Circuit and Signal Analysis, Digital and Control Systems, Electromechanics, Power Systems Energy Microprocessors and Programmable Controllers. Ability acquired by design project; both academic and personal experience.

EXPERIENCE

1985 - present **TRICHMAN ENGINEERING, INC.**
Control Design Engineer. Design software with logic program (ladder diagram logic relay) and hardware interface with programmable controller. Bailey network 90, Modicon 884 and Micro 84. Control AC motor (Speed Switch), reactor temperature (thermostat). damper (open and close), blower and conveyor system.

1984 - 1985 **U. S. RUBBER, INC.**
Control Design Engineer. Designed software with logic program (ladder diagram logic relay). Interfaced with design hardware for several applications (Analog Input, Analog Output), Programmable Controller (Allen-Bradley system) for industrial units (PLC 2/20, 2/15, 2/30 PLC 3). Motion control, torque control for DC/AC motors, temperature and pressure control used for tool machines.

1983 - 1984 **A. & Z. TORQUE SYSTEMS, INC.**
Product Engineer. Interfaced to manufacturing and customer, analog and digital systems involved, design, redesign, modification, DC motors, control systems, amplifier, power supply, encoder, decoder and pulse modulator.

1974 - 1978 **AMERICAN AIRPORT**
System Engineer. Tested equipment of radio receiver transmission, Electronic Communication, Analog and Digital System, Power System, Electromechanics, AC & DC motors, Dynamics, Alternator, and Generator.

REFERENCES

Available on request

Joan Clark
32 West 61st Street
New York, NY 10036
(212) 555-4814

Education

B.S.M.E, Pennsylvania State University, Scranton, PA, 1978
Ford Motor Company College Graduate Program, 1979-1980
California State Electrical Contractor License, Class C-10, 1988 to present

Employment Experience

January 1984 to April 1993
Oaneca Lake Vineyard, Arlsley, NY

ELECTROMECHANICAL ENGINEER, ELECTRICAL SUPERVISOR
Design, installation and maintenance of electric motor control stations for all wine processing equipment and vineyard pumping plants. Supporting installation and maintenance of winery H.V.A.C. and glycol cooling systems. All electrical and mechanical maintenance for supporting community buildings - restaurant, lounge, meeting hall, bath house, museum, and administrative offices (including computer hook-ups). Design and installation of several lighting and power systems for commercial and agricultural buildings. Familiar with security systems, satellite television systems, electronic troubleshooting, special lighting effects, tool fabrication, and drafting.

March 1979 to December 1983
General Motors Company, Dearborn, Michigan

PRODUCT DESIGN ENGINEER,
PRODUCT DEVELOPMENT ENGINEER
Design, development and testing of automotive driveline components.
Interplant liaison and supplier representative.

References

Available upon request

Joan Clark
32 West 61st Street
New York, NY 10036
(212) 555-4814

Education

B.S.M.E, Pennsylvania State University, Scranton, PA, 1978
Ford Motor Company College Graduate Program, 1979-1980
California State Electrical Contractor License, Class C-10, 1988 to present

Employment Experience

January 1984 to April 1993
Oaneca Lake Vineyard, Arlsley, NY

ELECTROMECHANICAL ENGINEER, ELECTRICAL SUPERVISOR
Design, installation and maintenance of electric motor control stations for all wine processing equipment and vineyard pumping plants. Supporting installation and maintenance of winery H.V.A.C. and glycol cooling systems. All electrical and mechanical maintenance for supporting community buildings - restaurant, lounge, meeting hall, bath house, museum, and administrative offices (including computer hook-ups). Design and installation of several lighting and power systems for commercial and agricultural buildings. Familiar with security systems, satellite television systems, electronic troubleshooting, special lighting effects, tool fabrication, and drafting.

March 1979 to December 1983
General Motors Company, Dearborn, Michigan

PRODUCT DESIGN ENGINEER,
PRODUCT DEVELOPMENT ENGINEER
Design, development and testing of automotive driveline components. Interplant liaison and supplier representative.

References

Available upon request

Field Support Engineer

JOSEPH D. MICHAELSON
452 N. Miners Avenue, #9
Brea, California 91760
(714) 555-1919

EXPERIENCE	**AMDAHL** Orange, California
3/89 to present	<u>FIELD ENGINEER</u>: Provide service activities for all models of Amdahl data processing systems including installation, discontinuance, relocation, diagnosis and repair. Responsible for trouble shooting and installation of engineering changes at customer sites.
	WILLIAMS DATA CENTER Tulsa, Oklahoma
4/86 to 2/89	<u>SENIOR COMPUTER OPERATOR</u>: Operated the computer console and unit record devices as lead operator and weekend manager. Coordinated work flow with scheduling, determined systems and programming failures, and initiated appropriate corrective action. Trained new and less experienced operators as required. The environment was IBM 3090, 3083, and AMDAHL V7 with MVS, IMS, CICS, TSO and RJE.
	AMERICAN AIRLINES Tulsa, Oklahoma
10/83 to 3/86	<u>COMPUTER OPERATOR</u>: Performed operator duties in an IBM 3083 environment with MVS and TSO.
	AMOCO-STANDARD OIL OF INDIANA Tulsa, Oklahoma
10/81 to 9/83	<u>COMPUTER OPERATOR</u>: Operated an IBM 4381 and 3081 with MVS, OS, and HASP. Also operated an IBM 1800 with TT 590 and 21 Track Tape Drive under DOS.
	SEISMOGRAPH SERVICE COMPANY Tulsa, Oklahoma
5/80 to 9/81	<u>LEAD COMPUTER OPERATOR/SHIFT SUPERVISOR</u>: Operated an IBM 4341 under DOS/VSE.
	UNITED STATES ARMY Camp Casey, Korea
9/78 to 2/80	<u>SHIFT SUPERVISOR</u>: Operated a UNIVAC 1008 computer system.

JOSEPH D. MICHAELSON
452 N. Miners Avenue, #9
Brea, California 91760
(714) 555-1919

EXPERIENCE   AMDAHL • Orange, California

3/89 to Present   ***FIELD ENGINEER***: Provide service activities for all models of Amdahl data processing systems including installation, discontinuance, relocation, diagnosis and repair. Responsible for trouble shooting and installation of engineering changes at customer sites.

    WILLIAMS DATA CENTER • Tulsa, Oklahoma

4/86 to 2/89   ***SENIOR COMPUTER OPERATOR***: Operated the computer console and unit record devices as lead operator and weekend manager. Coordinated work flow with scheduling, determined systems and programming failures, and initiated appropriate corrective action. Trained new and less experienced operators as required. The environment was IBM 3090, 3083, and AMDAHL V7 with MVS, IMS, CICS, TSO, and RJE.

    AMERICAN AIRLINES • Tulsa, Oklahoma

10/83 to 3/86   ***COMPUTER OPERATOR***: Performed operator duties in an IBM 3083 environment with MVS and TSO.

    AMOCO-STANDARD OIL OF INDIANA • Tulsa, Oklahoma

10/81 to 9/83   ***COMPUTER OPERATOR***: Operated an IBM 4381 and 3081 with MVS, OS, and HASP. Also operated an IBM 1800 with TT 590 and 21 Track Tape Drive under DOS.

    SEISMOGRAPH SERVICE COMPANY • Tulsa, Oklahoma

5/80 to 9/81   ***LEAD COMPUTER OPERATOR/SHIFT SUPERVISOR***: Operated an IBM 4341 under DOS/VSE.

    UNITED STATES ARMY • Camp Casey, Korea

9/78 to 2/80   ***SHIFT SUPERVISOR***: Operated a UNIVAC 1008 computer system.

Mechanical Engineer

Patricia Reed
11 Terrace Drive
South Windsor, CT 21456
(203) 666-9241
(203) 721-8000

CAREER OBJECTIVE

Position in mechanical engineering with emphasis on research and development as well as computer aided design or computer controlled systems.

EDUCATIONAL BACKGROUND

The University of Florida, Miami, Florida.
Bachelor of Science in Mechanical Engineering
Bachelor of Science in Materials Engineering, May 1984.
.

Emphasis of studies:
- Metallurgy
- Corrosion Engineering
- Welding
- Computer Science Software

- Electronics
- Dynamics
- Fatigue Analysis in Design
- Design

WORK EXPERIENCE

5/84–present

Mechanical Process Design Engineer
Power Systems Division/Fuel Cell Operations
Westinghouse Industries, Inc., Darien, CT.

"Elevated Temperature Tension-Torsion Fatigue Tests."
Research focuses on fatigue tests through the use of tension-torsion fatigue test machines. The study encompasses the use of computer software in evaluating the data generated. This program is headed by Dr. Eric H. Jordan, of the University of Connecticut, and is funded by NASA. (5/84-6/88)

Fatigue Analysis Researcher
Designed and detailed automated process/test equipment utilizing pneumatic, electric and mechanical actuators. Became familiar with ladder diagrams and relay logic. Performed tests to obtain analytical data needed for the design of vacuum chucks and part lifters. Acted as liaison between engineering, planning, vendors, and machine shop personnel. (7/88 – present)

OTHER ACTIVITIES

Member
American Society of Mechanical Engineers.
University of Connecticut Kyokushinkai Karate Club, President and Instructor.
Police Athletic League-Volunteer-Karate Instructor.

REFERENCES

Available upon request.

Patricia Reed

11 Terrace Drive
South Windsor, CT 21456
(203) 666-9241
(203) 721-8000

CAREER OBJECTIVE

A position in mechanical engineering with emphasis on research and development using Computer Aided Design (CAD) or Computer Controlled Systems.

EDUCATIONAL BACKGROUND

Bachelor of Science (BS), Mechanical Engineering
Bachelor of Science in Materials Engineering
University of Florida, 1984

Emphasis of Studies
- Metallurgy   - Electronics
- Corrosion Engineering   - Dynamics
- Welding   - Fatigue Analysis in Design
- Computer Science Software   - Design

WORK EXPERIENCE

POWER SYSTEMS DIVISION/FUEL CELL OPERATIONS
Westinghouse Industries, Inc., Darien, CT.

5/84-present

Mechanical Process Design Engineer: Designed and detailed automated process/test equipment utilizing pneumatic, electric and mechanical actuators. Familiar with ladder diagrams and relay logic. Performed tests to obtain analytic data needed for the design of vacuum chucks and part lifters. Acted as liaison between engineering, planning, vendors, and machine shop personnel. (7/88 - present)

Fatigue Analysis Researcher: *Elevated Temperature Tension-Torsion Fatigue Tests.* Research focuses on fatigue tests through the use of tension-torsion fatigue test machines. The study encompasses the use of computer software in evaluating the data generated. This program is headed by Dr. Eric H. Jordan, of the University of Connecticut, and is funded by NASA. (5/84 - 6/88)

OTHER ACTIVITIES

MEMBER
American Society of Mechanical Engineers
University of Connecticut Kyokushinkai Karate Club
        President and Instructor.
Police Athletic League -
        Volunteer and Karate Instructor.

REFERENCES

Available upon request

Mark Baldwin
15 Hudson Street
Bethesda, Maryland 41526
(301) 992-7251
(301) 765-7800

EDUCATION

University of Maryland at Baltimore, Maryland (School of Mines)
Bachelor of Science Degree, Nuclear Engineering May 1975
Professional Memberships: IEEE and ANS

SUMMARY

Eleven years experience in Nuclear Plant Engineering, including construction, start-up, modification, procedure writing, and outage maintenance. Also, installation and testing of ASME Code Class I, II, and III Systems and safety-related documents, including drawings, experience in QA documentation, scheduling, purchase memoranda and materials requisitions. Preparation of installation and acceptance test procedures, drawing updates, system walkoff, operability declaration, and documentation turnover associated with plant modifications.

PROFESSIONAL EXPERIENCE

The Virginia Valley Project   June 1986 to present
Seneca Nuclear Power Plant

I & C PROCEDURE WRITER: Responsible for review and rewrite of Instrumentation and Control Surveillance Instruction procedures. Areas included Foxboro and Bailey I & C loops, standby diesel generators, pressurizer pressure, response time testing and chlorine detection system. Procedures were written to assure compliance with Tech. Specs., FSAR, correct setpoints, ASME and IEEE standards, and vendor recommendations.

Maryland Power & Light Company (Contract)   November 1984 to April 1986
Brunswick Steam Electric Plant

SENIOR PLANT MODIFICATION ENGINEER: Responsible for the review, approval, implementation and closeout of mechanical, instrumentation and electrical modifications. Responsibilities included scope and engineering, scheduling, preparation of installation and acceptance test procedures, final walkout, documentation turnover, and operability. Also responsible for updating the FSAR, and assuring availability of spare parts and vendor manuals.

Chambers Power Corporation   February 1976 to November 1984
Attendo Nuclear Power Project

<u>SENIOR MECHANICAL ENGINEER</u>: Responsible for approving and signing safety related documents for compliance with applicable codes, standards and procedures. Applicable drawings include Isometric, Plumbing, P&ID's, design criteria, specifications, purchase memoranda and material requisitions. Instructed training classes on and wrote project procedures. Interfaced with various organizations performing project audits, including QA, NRC and ASME.

REFERENCES

Furnished upon request.

Mark Baldwin

15 Hudson Street
Bethesda, Maryland 41526
(301) 992-7251
(301) 765-7800

EDUCATION

University of Maryland at Baltimore, Maryland (School of Mines)
Bachelor of Science Degree, Nuclear Engineering May 1975
Professional Memberships: IEEE and ANS

SUMMARY

Eleven years experience in Nuclear Plant Engineering, including construction, start-up, modification, procedure writing, and outage maintenance. Also, installation and testing of ASME Code Class I, II, and III Systems and safety-related documents, including drawings, experience in QA documentation, scheduling, purchase memoranda and materials requisitions. Preparation of installation and acceptance test procedures, drawing updates, system walkoff, operability declaration, and documentation turnover associated with plant modifications.

PROFESSIONAL EXPERIENCE

The Virginia Valley Project   **June 1986 to present**
Seneca Nuclear Power Plant

I & C PROCEDURE WRITER: Responsible for review and rewrite of Instrumentation and Control Surveillance Instruction procedures. Areas included Foxboro and Bailey I & C loops, standby diesel generators, pressurizer pressure, response time testing and chlorine detection system. Procedures were written to assure compliance with Tech. Specs., FSAR, correct setpoints, ASME and IEEE standards, and vendor recommendations.

Maryland Power & Light Company (Contract) November 1984 to April 1986
Brunswick Steam Electric Plant

SENIOR PLANT MODIFICATION ENGINEER: Responsible for the review, approval, implementation and closeout of mechanical, instrumentation and electrical modifications. Responsibilities included scope and engineering, scheduling, preparation of installation and acceptance test procedures, final walkout, documentation turnover, and operability. Also responsible for updating the FSAR, and assuring availability of spare parts and vendor manuals.

Chambers Power Corporation   February 1976 to November 1984
Attendo Nuclear Power Project

SENIOR MECHANICAL ENGINEER: Responsible for approving and signing safety related documents for compliance with applicable codes, standards and procedures. Applicable drawings include Isometric, Plumbing, P&ID's, design criteria, specifications, purchase memoranda and material requisitions. Instructed training classes on and wrote project procedures. Interfaced with various organizations performing project audits, including QA, NRC and ASME.

REFERENCES

Furnished upon request.

Scientist

Ronald Evanridge
123 Rockwater Avenue
Ridgecrest, California 91426
(818) 555-6445

WORK EXPERIENCE:

Sept. 1989 - Jan. 1993   Cantrell Engineering Services Group
Riverdale, California
Chief Scientist
Advanced Technology and Research

Responsibilities

System design of optics for Sidewinder modification.
System design of optics for Focal Plane Array.
Research in the thermal physics of optical thin film coating for the Advanced Space-Based
Laser Platform.
Development of mathematical physics to describe thermal transient phenomena of single
and repetitive pulsed high energy laser optics. Development of Maxwell's equation to
describe field vectors in optical thin film coatings.
Research in vulnerability of optical sensor, seekers, and optical trains to CW, single, and
repetitive pulsed laser damage. Development of mathematical model to describe
destruction by lasers of optical capability for Sidewinder, Maverick, Hellfire, TOW,
Tomahawk, LGB, FLIR, LLTV, and advanced optical systems on satellite platforms.
Research in generic laser damage to optical components.

Feb. 1983 - Sept. 1989   Arco Engineering
Hollywood, California
Senior Staff Consultant

Responsibilities

Research in generic laser damage to optical components.
Development of mathematical model to describe optical component and field distortion for
various types of high energy laser systems.
Research vulnerability of sensors and optical trains to high energy laser attack.
Development of thermodynamic model to describe transient heat transfer and fluid
mechanics of cooled laser mirrors.
Development of mathematical model to describe the optical distortion, defect, and
catastrophic damage limitations for MIRACL, NACL, ROPCL, COIL, ALPHA, and
EMRLD chemical laser optical components.

SECURITY CLASSIFICATION   Active Secret

EDUCATION

Ph.D., Physical Chemistry, University of Michigan, Ann Arbor, Mich., 1979
M.S., Chemical Engineering, Michigan State College, Ann Arbor, Mich., 1973
B.S., Chemical Engineering, University of Florida, Miami, Fla., 1969
A.A., Mechanical Engineering, Long Beach City College, Long Beach, Calif., 1966

HONORS AND AWARDS

1989 Recipient of International Society for Optical Engineering's
    Isidore Baney Medal and Prize.
Who's Who in the West, 20th ed., 1991 and 21st ed., 1992
International Who's Who in Optical Science and Engineering, 1991
Who's Who in America, 1992
Who's Who in the World, 1992
Member:
- Sigma Xi Honor Society
- S.P.I.E.
- Optical Society of America
- A.A.A.S
- New York Academy of Sciences

PROFESSIONAL PUBLICATIONS

BOOKS

Laser Reflection in Solution, Macmillan Publishing Co., New York, NY, 1992

Co-Author - *Solvent Jet Technology,* Technology Utilization Service,
National Aeronautical and Space Administration, Washington, DC,
NASA Contract NASA SP-5033, 1972.

ARBG Contributions to Metal Joining, Technology Utilization Service,
National Aeronautical and Space Administration, Washington, DC,
NASA Contract NASA SP-5064, 1973.

Gaseous Measurement and Calibration, Technology Utilization Service,
National Aeronautical and Space Administration, Washington, DC,
NASA Contract NASA-1517, SP-5033, 1974.

REFERENCES   Available upon request.

Ronald Evanridge
123 Rockwater Avenue
Ridgecrest, California 91426
(818) 555-6445

WORK EXPERIENCE:

Sept. 1989 - Jan. 1993
Cantrell Engineering Services Group
Riverdale, California
Chief Scientist
Advanced Technology and Research

Responsibilities

System design of optics for Sidewinder modification.
System design of optics for Focal Plane Array.
Research in the thermal physics of optical thin film coating for the Advanced Space-Based Laser Platform.
Development of mathematical physics to describe thermal transient phenomena of single and repetitive pulsed high energy laser optics. Development of Maxwell's equation to describe field vectors in optical thin film coatings.
Research in vulnerability of optical sensor, seekers, and optical trains to CW, single, and repetitive pulsed laser damage. Development of mathematical model to describe destruction by lasers of optical capability for Sidewinder, Maverick, Hellfire, TOW, Tomahawk, LGB, FLIR, LLTV, and advanced optical systems on satellite platforms.
Research in generic laser damage to optical components.

Feb. 1983 - Sept. 1989
Arco Engineering
Hollywood, California
Senior Staff Consultant

Responsibilities

Research in generic laser damage to optical components.
Development of mathematical model to describe optical component and field distortion for various types of high energy laser systems.
Research vulnerability of sensors and optical trains to high energy laser attack.
Development of thermodynamic model to describe transient heat transfer and fluid mechanics of cooled laser mirrors.
Development of mathematical model to describe the optical distortion, defect, and catastrophic damage limitations for MIRACL, NACL, ROPCL, COIL, ALPHA, and EMRLD chemical laser optical components.

SECURITY CLASSIFICATION   Active Secret

EDUCATION

Ph.D., Physical Chemistry, University of Michigan, Ann Arbor, Mich., 1979
M.S., Chemical Engineering, Michigan State College, Ann Arbor, Mich., 1973
B.S., Chemical Engineering, University of Florida, Miami, Fla., 1969
A.A., Mechanical Engineering, Long Beach City College, Long Beach, Calif., 1966

HONORS AND AWARDS

1989 Recipient of International Society for Optical Engineering's
   Isidore Baney Medal and Prize.
Who's Who in the West, 20th ed., 1991 and 21st ed., 1992
International Who's Who in Optical Science and Engineering, 1991
Who's Who in America, 1992
Who's Who in the World, 1992
Member:
   - Sigma Xi Honor Society
   - S.P.I.E.
   - Optical Society of America
   - A.A.A.S
   - New York Academy of Sciences

PROFESSIONAL PUBLICATIONS

BOOKS

Laser Reflection in Solution, Macmillan Publishing Co., New York, 1992

Co-Author - *Solvent Jet Technology,* Technology Utilization Service,
National Aeronautical and Space Administration, Washington, DC,
NASA Contract NASA SP-5033, 1972.

ARBG Contributions to Metal Joining, Technology Utilization Service,
National Aeronautical and Space Administration, Washington, DC,
NASA Contract NASA SP-5064, 1973.

Gaseous Measurement and Calibration, Technology Utilization Service,
National Aeronautical and Space Administration, Washington, DC,
NASA Contract NASA-1517, SP-5033, 1974.

REFERENCES   **Available upon request.**

MICHAEL J. HARDING
12332 Sherry Lane
Garden Grove, CA 92540

(714) 555-9753

WORK EXPERIENCE

SECURITY PACIFIC BANK   Dec 89 to Sep 92
Seattle, Washington

SENIOR PROGRAMMER/ANALYST: Designed programs and maintained systems on Dun & Bradstreet (Payroll Package) and Credit Card
Generation System. Implemented McCormack & Dodge Payroll System
and rewrite of the General Ledger System. Wrote numerous standalone COBOL programs to process Oregon state taxes. Performed rewrite and update of the United Way System from DYL-280 to COBOL. In
charge of all systems relating to credit card systems and credit card
data bases. Systems environment was IBM 3083 with IMS/DB/DC, DB2,
CICS, COBOL, and COBOL II.

AUTO CLUB OF SOUTHERN CALIFORNIA   Aug 86 to Dec 89
Costa Mesa, California

SYSTEMS ANALYST II: In the IBM 3083k Mainframe environment, developed, programmed, and implemented Card Elimination System. Project lead on hard copy to microfiche conversion. Developed the integrated budget system in an IBM Microcomputer environment using
APPLICATION SYSTEM (4GL Language from IBM). Also served as a PC
and Mainframe Repair Specialist. Project involvement was from initial analysis and planning, to systems analysis and design, and completion through coding, testing, and user approval. Systems environment was IBM 3083k Mainframe and IBM PC/XT/AT/386/PS2
Microcomputers. Software environment included CICS/PL1, PL1, COBOL,
SAS, AS, DATAGRAPHIX, COMPAREX, SDF, ALC, plus numerous PC
software packages.

NORTHROP CORPORATION   May 85 to Aug 86
Advanced Systems Division
Pico Rivera, California

SENIOR COMPUTER TRAINER/PROGRAMMER ANALYST: In an IBM
Mainframe, HP-3000, and IBM Microcomputer environment, developed
and instructed classes in the commercial software packages used by
Northrop. Also performed projects for Area Managers requiring additional Information Systems resources and where special knowledge
and skills were required.

Projects included production scheduling, stock inventory & procurement, structural variance, accounting, and terrain guidance. Was Northrop's only IMS/DB/DC Instructor, designed and set up their training system for that product. Only Instructor trained on the HP-3000 System using IMAGE Database and POWERHOUSE Software. Project Leader for the *Resume Tracking System* on the HP-3000.

TRW CREDIT CORPORATION   Aug 84 to May 85
Orange, California

PROGRAMMER II: Designed, developed, and programmed a variety of applications in the IBM Mainframe and PC environments. Used COBOL for projects ranging from business credit validation to electronic mail. Performed implementation and repair of CICS and DIALOG MANAGER screens for security and data transmission processing. System environment was IBM, 30XX, CICS, COBOL, ALC, EASYTRIEVE, and various PC software packages.

UNITED STATES MARINE CORPS   Aug 67 to Jul 84
El Toro, California

SENIOR PROGRAMMER/DATA PROCESSING SHOP CHIEF: Designed and programmed a variety of projects which included a graphics display for a battlefield computer, a flight readiness data and aircraft maintenance scheduling system, a parts inventory system, a personnel system, and a payroll processing system. The systems environment was IBM Mainframe and PC with programming in COBOL.

EDUCATION

University of Washington, Data Base Design, Seattle, Washington, 1991-1992

Orange Coast Community College, Computer Science, Costa Mesa, California, 1986-1989

Santiago Community College, Computer Science, Santa Ana, California, 1980-1986

University of Washington, Engineering, Seattle, Washington, 1973-1974

Wenatchee Valley College, General Subjects, Wenatchee, Washington, 1972-1973

MICHAEL J. HARDING
12332 Sherry Lane
Garden Grove, CA 92640

(714) 555-9753

WORK EXPERIENCE

SECURITY PACIFIC BANK   **Dec 89 to Sep 92**
Seattle, Washington

SENIOR PROGRAMMER/ANALYST: Designed programs and maintained systems on *Dun & Bradstreet (Payroll Package)* and *Credit Card Generation System*. Implemented *McCormack & Dodge Payroll System* and rewrite of the *General Ledger System*. Wrote numerous standalone COBOL programs to process *Oregon State Taxes*. Performed rewrite and update of the *United Way System* from DYL-280 to COBOL. In charge of all systems relating to credit card systems and credit card data bases. Systems environment was IBM 3083 with IMS/DB/DC, DB2, CICS, COBOL, and COBOL II.

AUTO CLUB OF SOUTHERN CALIFORNIA   **Aug 86 to Dec 89**
Costa Mesa, California

SYSTEMS ANALYST II: In the IBM 3083k Mainframe environment, developed, programmed and implemented *Card Elimination System*. Project lead on hard copy to microfiche conversion. Developed the *Integrated Budget System* in an IBM Microcomputer environment using APPLICATION SYSTEM (4GL Language from IBM). Also served as a PC and Mainframe Repair Specialist. Project involvement was from initial analysis and planning, to systems analysis and design, and completion through coding, testing, and user approval. Systems environment was IBM 3083k Mainframe and IBM PC/XT/AT/386/PS2 Microcomputers. Software environment included CICS/PL1, PL1, COBOL, SAS, AS, DATAGRAPHIX, COMPAREX, SDF, ALC, plus numerous PC software packages.

NORTHROP CORPORATION   **May 85 to Aug 86**
Advanced Systems Division
Pico Rivera, California

SENIOR COMPUTER TRAINER / PROGRAMMER ANALYST: In an IBM Mainframe, HP-3000, and IBM Microcomputer environment, developed and instructed classes in the commercial software packages used by Northrop. Also performed projects for Area Managers requiring additional Information Systems resources and where special knowledge and skills were required. Projects included production scheduling, stock inventory & procurement, structural variance, accounting, and terrain guidance.

NORTHROP CORPORATION (continued)

As Northrop's only IMS/DB/DC Instructor, designed and set up their training system for that product. Only Instructor trained on the HP-3000 System using IMAGE Database and POWERHOUSE Software. Project Leader for the *Resume Tracking System* on the HP-3000.

TRW CREDIT CORPORATION   Aug 84 -May 85
Orange, California

PROGRAMMER II: Designed, developed, and programmed a variety of applications in the IBM Mainframe and PC environments. Used COBOL for projects ranging from business credit validation to electronic mail. Performed implementation and repair of CICS and DIALOG MANAGER screens for security and data transmission processing. System environment was IBM 30XX, CICS, COBOL, ALC, EASYTRIEVE, and various PC software packages.

UNITED STATES MARINE CORPS   Aug 67 to Jul 84
El Toro, California

SENIOR PROGRAMMER / DATA PROCESSING SHOP CHIEF: Designed and programmed a variety of projects which included a graphics display for a battlefield computer, a flight readiness data and aircraft maintenance scheduling system, a parts inventory system, a personnel system, and a payroll processing system. The systems environment was IBM Mainframe and PC with programming in COBOL.

University of Washington, Data Base Design, Seattle, Washington, 1991-1992

Orange Coast Community College, Computer Science, Costa Mesa, California, 1986-1989

Santiago Community College, Computer Science, Santa Ana, California, 1980-1986

University of Washington, Engineering, Seattle, Washington, 1973-1974

Wenatchee Valley College, General Subjects, Wenatchee, Washington, 1972-1973

Senior Electronic Engineer

ALICE SPRING
1600 Sprague Street
Cleveland, OH 45678
(219) 813-4416 (Home)

OBJECTIVE   SENIOR ELECTRONIC ENGINEER

EDUCATION   MASTER OF SCIENCE IN ELECTRICAL ENGINEERING
Ohio State University, Akron, Ohio, 1981
Took nine courses in Digital Engineering

BACHELOR OF SCIENCE IN ELECTRICAL ENGINEERING
University of Science and Industry, Tehran, Iran, 1977

Electrical Engineering Courses
California State University at Los Angeles, 1979

EXPERIENCE   United Technology, Inc.; Cleveland, Ohio

May 1982 to
Present

SENIOR DIGITAL DESIGN ENGINEER
With minimum of supervision carried out all steps to
complete four projects, including architect, timing analysis,
circuit design, writing PAL equations, debugging prototype
designs, preparing pin lists and programs in assembly
language for project development and manufacturing.
Prepared product manuals and schematic diagrams for
internal use and for technical writers. Supervised the work
of engineering assistant. Performed at a senior level and
received excellent salary raises and appreciation.

Campus Information Systems; Ohio State University, Akron,
Ohio

Jan 1980 to
Mar 1982

ELECTRONIC TECHNICIAN
Install, troubleshoot, and maintain campus computer
equipment. Supervise other students' work. Provide
technical services to many departments. When supervisor
left, was asked to take his job.

SKILLS   Expert in 68000 CPU Digital Equipment Co. LSI bus (Q-bus).
Know 6502, 6800, 8085, 8086/8088, 80386, 80486, 68000
assembly language; Motorola VME bus. Familiar with UNIX,
MS/DOS, CP/M, Windows, OS/2 C++ Language; FORTRAN,
PASCAL. Since I terminated my last employment, I have
studied Software and Hardware of microcomputers and have
gained strong knowledge in area. Possess management
ability. Above average understanding of issues in mechanics.

PERSONAL   Bilingual in English and Farsi

REFERENCES   Available on request

Senior Electronic Engineer (Enhanced)

ALICE SPRING
1600 Sprague Street
Cleveland, OH 45678
(219) 813-4416 (Home)

OBJECTIVE   **SENIOR ELECTRONIC ENGINEER**

EDUCATION   **MASTER OF SCIENCE IN ELECTRICAL ENGINEERING**
Ohio State University, Akron, Ohio, 1981
Took nine courses in Digital Engineering

BACHELOR OF SCIENCE IN ELECTRICAL ENGINEERING
University of Science and Industry, Tehran, Iran, 1977

Electrical Engineering Courses
California State University at Los Angeles, 1979

EXPERIENCE   United Technology, Inc.; Cleveland, Ohio

May 1982 to   SENIOR DIGITAL DESIGN ENGINEER
Present   With minimum of supervision carried out all steps to complete four
projects, including architect, timing analysis, circuit design, writing
PAL equations, debugging prototype designs, preparing pin lists and
programs in assembly language for project development and
manufacturing. Prepared product manuals and schematic diagrams for
internal use and for technical writers. Supervised the work of
engineering assistant. Performed at a senior level and received
excellent salary raises and appreciation.

Campus Information Systems; Ohio State University, Akron, Ohio

Jan 1980 to   ELECTRONIC TECHNICIAN
Mar 1982   Install, troubleshoot, and maintain campus computer equipment.
Supervise other students' work. Provide technical services to many
departments. When supervisor left, was asked to take his job.

SKILLS   Expert in 68000 CPU Digital Equipment Co. LSI bus (Q-bus). Know
6502, 6800, 8085, 8086/8088, 80386, 80486, 68000 assembly language;
Motorola VME bus. Familiar with UNIX, MS/DOS, CP/M, Windows,
OS/2 C++ Language; FORTRAN, PASCAL. Since I terminated my last
employment, I have studied Software and Hardware of microcomputers
and have gained strong knowledge in area. Possess management ability.
Above average understanding of issues in mechanics.

PERSONAL   Bilingual in English and Farsi

REFERENCES   Available on request

Structural Engineer

William Brooks
120 East 10th Street
Scranton, PA 10082
(219) 555-3221

OBJECTIVE   Looking for challenging job to work as a Structural Engineer.

PROFESSIONAL EXPERIENCE

Oct 1990   Universal Instruments, Inc.
to   Bombay, India
March 1993

                  Structural Engineer. Extensive experience in the Analysis and Design of
                  steel/reinforced concrete structures for static (Dead load, live load,
                  Thermal load) or dynamic (Wind, Earthquake) loading from foundation to
                  the superstructure including plastic design of multistory braced frames.
                  Thorough knowledge of different design procedure based on hand
                  calculations such as a slope- deflection method, moment-distribution
                  method, ultimate load theory, matrix displacement method and plastic
                  design.

Jan. 1988   Pace Engineering
to   Baroda, India
Sept. 1990

                  Civil Engineer. Analysis and design of steel and reinforced concrete
                  structures for static or dynamic loading from foundation to the super
                  structure including plastic design of multistory braced frames with
                  implementation of computer application.

EDUCATIONAL BACKGROUND

Aug. 1990   MS in Civil Engineering
to   University of Bombay, Bombay, India
Dec. 1992   Passed First Class with Distinction (Rank 2nd out of 45)

Major: Structural Engineering

Thesis: Study of Stress intensity factor for cement mortar
                  beams with different initial notch lengths, containing
                  one percent of steel fibers and the beams tested under
                  three point loading system.

Course Work   Finite element methods, Matrix analysis of framed structures, Theory
                  of plates and shell, Advanced steel design, Advanced concrete design.

June 1988 to March 1990	BS in Civil Engineering University of Bombay, Bombay, India
Major	Structural Engineering
Thesis	Applications of Energy theorems
AFFILIATIONS	Member of AICE.
HOBBIES	Swimming, Reading, Gardening, and Travel
PERSONAL INFORMATION	Permanent Resident Ready to relocate and travel
REFERENCES	Will be furnished upon request

William Brooks

120 East 10th Street
Scranton, PA 10082
(219) 555-3221

OBJECTIVE   Looking for challenging job to work as a Structural Engineer.

PROFESSIONAL EXPERIENCE:

Oct 1990   **Universal Instruments, Inc.**
to   **Bombay, India**
March 1993

<u>Structural Engineer</u>. Extensive experience in the Analysis and Design of steel/reinforced concrete structures for static (Dead load, live load, Thermal load) or dynamic (Wind, Earthquake) loading from foundation to the superstructure including plastic design of multistory braced frames. Thorough knowledge of different design procedure based on hand calculations such as a slope- deflection method, moment-distribution method, ultimate load theory, matrix displacement method and plastic design.

Jan. 1988   **Pace Engineering**
to   **Baroda, India**
Sept. 1990

<u>Civil Engineer</u>. Analysis and design of steel and reinforced concrete structures for static or dynamic loading from foundation to the super structure including plastic design of multistory braced frames with implementation of computer application.

EDUCATIONAL BACKGROUND

Aug. 1990   **MS in Civil Engineering**
to   University of Bombay, Bombay, India
Dec. 1992   Passed First Class with Distinction (Rank 2nd out of 45)

Major: Structural Engineering

Thesis: Study of Stress intensity factor for cement mortar
                beams with different initial notch lengths, containing
                one percent of steel fibers and the beams tested under
                three point loading system.

Course Work   Finite element methods, Matrix analysis of framed structures, Theory
                of plates and shell, Advanced steel design, Advanced concrete design.

June 1988 to March 1990	**BS in Civil Engineering** University of Bombay, Bombay, India
Major	Structural Engineering
Thesis	Applications of Energy theorems

AFFILIATIONS Member of AICE.

HOBBIES Swimming, Reading, Gardening, and Travel

PERSONAL Permanent Resident
INFORMATION Ready to relocate and travel

REFERENCES Will be furnished upon request

Technical Support Manager (Bullet)

WILLIAM A. PETERSEN
4767 Willow Way
New Castle, Indiana 46987
(317) 555-4637

A Senior Manager with over 20 years of experience in the computer industry, the last 10 years with Wang Laboratories. Responsibilities were managing and directing support and providing consulting services for Regional and Field Service.

MANAGERIAL:

* Directed staff of 5 managers and 39 analysts in sales support and client consulting services and produced over $1,5000,000 in revenue the first year.
* Formulated strategies and implemented plans to help close sales opportunities resulting in the sale of multiple data processing and imaging systems.
* Organized and directed staff of 12 technical specialists to support Field Service District of over 60 Customer Engineers.
* Increased customer satisfaction by resolving and reducing critical field escalations by 25%.
* Led task force in analyzing scope and responsibilities of Regional Business Services Group with recommendations to VP for redefinition of practices and increased effectiveness.
* Planned and directed support of Imaging Showcase involving over 25 software vendors.
* Planned and implemented consolidation of two Field Service Districts resulting in increased production with fewer personnel.

TECHNICAL:

* Participated with sales and support staff in formulating responses to RFP's and RFQ's on LAN and imaging systems in excess of $1,000,000.
* Conducted quality assurance reviews of client orders.
* Project managed implementation of voice response system for major insurance client utilizing formal project management tools and techniques.
* Developed and wrote application handbook for use with project management software.

ADMINISTRATIVE:

* Developed monthly consulting and revenue forecasts.
* Analyzed operational statistics and made recommendations for productivity improvements.
* Assisted in development of district budget.

WORK HISTORY: 1983 to Present   WANG LABORATORIES INC.

1991-Present   Regional Sales Support Manager/Project Manager
1987-1991   District Technical Support Manager
1983-1987   Area Operations Manager/Branch Manager

1973-1983   GTE Information Systems — Design Engineer

EDUCATION:

MBA, Marketing Management, Webster University, 1982
BS, Industrial Engineering, Purdue University, 1973

WILLIAM A. PETERSEN
4767 Willow Way
New Castle, Indiana 46987
(317) 555-4637

A Senior Manager with over 20 years of experience in the computer industry, the last 10 years with Wang Laboratories. Responsibilities were managing and directing support and providing consulting services for Regional and Field Service.

MANAGERIAL:

- Directed staff of 5 managers and 39 analysts in sales support and client consulting services and produced over $1,500,000 in revenue the first year.
- Formulated strategies and implemented plans to help close sales opportunities resulting in the sale of multiple data processing and imaging systems.
- Organized and directed staff of 12 technical specialists to support Field Service District of over 60 Customer Engineers.
- Increased customer satisfaction by resolving and reducing critical field escalations by 25%.
- Led task force in analyzing scope and responsibilities of Regional Business Services Group with recommendations to VP for redefinition of practices and increased effectiveness.
- Planned and directed support of Imaging Showcase involving over 25 software vendors.
- Planned and implemented consolidation of two Field Service Districts resulting in increased production with fewer personnel.

TECHNICAL:

- Participated with sales and support staff in formulating responses to RFP's RFQ's on LAN and imaging systems in excess of $1,000,000.
- Conducted quality assurance reviews of client orders.
- Project Managed implementation of voice response system for major insurance client utilizing formal project management tools and techniques.
- Developed and wrote application handbook for use with project management software.

ADMINISTRATIVE:

- Developed monthly consulting and revenue forecasts.
- Analyzed operational statistics and made recommendations for productivity improvements.
- Assisted in development of district budget.

WORK HISTORY: 1983 to Present   WANG LABORATORIES INC.

1991 - Present	Regional Sales Support Manager / Project Manager
1987 - 1991	District Technical Support Manager
1983 - 1987	Area Operations Manager / Branch Manager
1973 - 1983	**GTE Information Systems** -- Design Engineer

EDUCATION:

MBA, Marketing Management, Webster University, 1982
BS, Industrial Engineering, Purdue University, 1973

Section 2

Resumes That Got Interviews

The resumes on the following pages are produced as they were received from the people who created them. Although some of them date back a number of years, they remain good examples of effective resumes. Only the names, addresses, and phone numbers have been changed.

These resumes are good examples of effective resumes because each of them has been instrumental in getting interviews for their owners. While they are excellent models, they are also unique. As you examine these resumes, you should be aware of two things.

First, they represent a variety of resume styles, types, and hybrids. This means that there is no best format or way to prepare a resume. As you review them, place yourself in the place of a prospective employer and ask yourself, "Why would I want (or not want) to interview this candidate?" It is from this perspective that you must create your resume.

Second, observe that many of the resumes do not follow the conventional approaches to resume writing. Most do not even follow the advice given in this book and might even be labeled as maverick in their approach. Several violate the "one page" rule and use two pages. You should also know that many resumes that were more than two pages long were considered for inclusion in this book but were omitted because of space limitations. Others include information that might be sensitive (personal) or questionable ("Why was that item included?"). The bottom line is that these resumes reached employers and did the job, that is, they got interviews.

JOHN L. REID
35791 Calle de Papel • Dana Point, California 92799
(714) 555-0293

EDUCATION

M.B.A. Candidate at University of Phoenix, Arizona—Graduation, October 1991

S.SC., Deans List, California State University at Long Beach
Major: Finance   Options: Investments and Financial Management

NASD Licenses: Series 7-General Securities, Series 3-Futures, Series 63-Interstate Rule
144 Restricted Securities Registered

Insurance Licenses: California—Life, Disability, Variable Annuity, Fire, and Casualty

BUSINESS EXPERIENCE

Twenty years in marketing, finance, and sales. Also program management in an aerospace manufacturing environment. Successfully dealt with turnaround situations and functioned at a management level. Resided in Great Britain, five years; Canada, seven years; travelled throughout Western Europe and South Africa; culturally knowledgeable. Areas of expertise:

- Development of Marketing Strategies & Sales
- Investment Portfolio & Money Management
- Program Management (Manufacturing)
- Business Development & Commercial Lending
- Presentations & Proposal Writing
- Microcomputer Systems & Software
- Risk Analysis & Underwriting
- Negotiation of Major Contracts

ACCOMPLISHMENTS

- Built a successful business in the securities industry from a zero base to assets over $10 million, cash over $6 million, and a client base exceeding 700 accounts.
- Developed for a manufacturing concern turnkey computer forecasting and risk analysis programs, which employed discounted cash flow analysis.
- Successfully negotiated multimillion dollar aerospace contracts and captured new business with innovative proposals. Provided management consulting with emphasis on financial structure and cash flow.
- Top district salesman, first QTR 1971; sale of electronic components and equipment.

EMPLOYMENT HISTORY

June 1984–Present	Dean Witter Reynolds Inc. • Laguna Hills & Orange, CA *Account Executive*
May 1980–May 1984	Parker Hannifin Aerospace Hydraulic Division • Irvine, CA *Senior Contract Administrator/Program Manager*
Oct. 1979–April 1980	Aerojet General Valve Co. • Fullerton, CA *Contracts Administrator*
Nov. 1978–Sept.1979	Swedlow Inc. • Garden Grove, CA *Contract Administrator*
July 1977–Oct.1978	Home Bank • Signal Hill, CA *Installment & Commercial Lending Officer*
May 1969–Oct. 1973	Tandy Corp. • Long Beach CA *Salesman/Assistant Manager*

MONICA STEWART SOTO
26677 Tricolor Canyon Drive
Bellflower, California 92647
(714) 555-3758

EDUCATION   BS, Marine Biology,
California State University at Long Beach, 1977

EXPERIENCE   FLUOR CORPORATION • Irvine, CA

3/82 to 8/82   *Associate Systems Analyst*: Responsible for 130 COBOL programs in the administrative area including an Applicant Flow System which utilizes Command Level CICS. In process of converting the Executive Compensation System to IMS and have attended 3 IMS classes in support of this effort. Environment is IBM 3033 and 3081 under MVS. Software support includes CICS, IMS, ANSWERDB, ANSWER2, ASI-ST, LIBRARIAN, and TSO with FSE and SPF. All application programming is in COBOL using structured methodology.

GTE DATA SERVICES • Marina Del Rey, CA

10/79 to 3/82   *Programmer/Analyst*: Developed and maintained COBOL programs in the area of inventory control. Developed, programmed, implemented, documented, and maintained a material cost tracking called Carrier Tracking. Interfaced with financial systems users for the development of program specifications. Started at GTE Data Services as a Programmer Trainee and completed a 6 month formal training course in OS/JCL and COBOL. The systems environment was (2) IBM 3033's, (1) AMDAHL 470V6 under MVS. Software included CICS, DYL-260, ISAM and VSAM access methods, TMS, PANVALET, and TSO/SPF. Programming was in COBOL with approximately 40% development and 60% maintenance.

CONTINENTAL AIRLINES • Playa del Rey, CA

10/77 to 10/79   Check-In Specialist: Assisted Flight attendants in checking in and out for scheduled flights.

PERSONAL   Enjoy travel

JOHN V. PARKER
18727 Madrid Circle
Westminster, CA 92677
(714) 555-9564

MIS Management
Systems Installation/Client Services

A seasoned professional with a comprehensive background in systems, methods, procedures, and controls in a sophisticated computer environment. Exceptional capabilities in planning, organizing, and managing large scale computer system projects including installation, support, and coordination. Functional areas of involvement have included administration, operations, production, and sales support.

CAREER RETROSPECTIVE

Lake Charles Memorial Hospital

1988–1990   *Chief Information Officer* • Lake Charles, Louisiana

Manage all aspects of the hospital's information services functions. Supervise the work product of 20 professionals involved in Information Systems, the ROLM CBX system, and personal computers. Specific responsibilities have included installing IBM AS/400 and migration of the software from the IBM System/38, upgrading the ROLM 7000 to the ROLM 8000, and installing numerous new software applications.

- Formulated and developed an Information Systems Master Plan to guide the hospital's future actions regarding systems decisions.

- Implemented a User Request system to track the progress on modifications and inform user personnel of priorities and time frames.

Pacific Health Resources

1982-1988   *Director* • Installations/Client Services, Los Angeles, California.

Governed the diverse aspects of installations, client services, and manpower development for this service organization. Presided over a staff of nine that installed 16 systems and supported over 40 production applications with user clients. Specific responsibilities included marketing products to existing client base, developing system proposals, negotiating software contracts, coordinating system implementations, and coordinating all client services for system support.

- Wrote procedures and outlined all job functions relative to coordinating all activity for systems between the client facilities and the data center.

- Developed a personal computer consulting group that offered supporting services for the acquisition of hardware/software, programming and set-up, and technical training.

- Provided marketing support for software sales at conventions and client sites by demonstrating software and discussing its technical applications.

Price Waterhouse

1981–1982   *Manager* • EDP Healthcare Consulting Group, St. Louis, Missouri.

Managed the EDP services in a healthcare group of seven people. Responsibilities included practice development, personnel development and work performance.

■ Worked closely with the senior manager in establishing a healthcare practice in St. Louis for local clients.

■ Involved in two major engagements, vendor selection and implementation assistance for one client, and a comprehensive requirements definition study (RFP development, vendor proposal evaluation, system selection) with the other client.

Earlier experience has included progressively responsible operations/installation positions with the following:

Sisters of Charity Hospital Corp., Houston, Texas
Corporate Systems Coordinator (1979–1981) SCH/PCS Products

Shared Medical Systems, Los Angeles, California
Installation Director (1977–1979) On-line and Batch Systems

Hoag Memorial Hospital, Newport Beach, California
Computer Operations Manager (1969–1977)

PROFESSIONAL AFFILIATIONS

Data Processing Management Association (DPMA)

Electronic Computing Health Oriented (ECHO)

Baxter Delta User Group

EDUCATION

1976 Bachelor of Business Administration
California State University, Fullerton, California

ROBERT A. BARBENEAU
12765 Gary Street
Northridge, California 91200
(818) 555-1992

EXPERIENCE   Headquarters, United States Marine Corps
                   Washington, D.C.

Feb 82 to Present ***Network Technician***. Serving as a member of a Network
                   Control Team. Responsible for installing and testing
                   teleprocessing peripheral devices, locating and identifying
                   the cause of teleprocessing systems failure, and entering
                   correct commands to control the teleprocessing network.
                   Tasks also involve monitoring the network's performance,
                   operating and maintaining a diagnostic modem network, and
                   performing first level corrective maintenance for related
                   teleprocessing hardware and software problems. Network
                   configuration, installation, and maintenance included the
                   laying of both coaxial and four wire cable, installation of
                   controllers, terminals, printers, modems, and troubleshooting
                   the data communication network utilizing halcyon test
                   equipment and patch panels.

                   Software: OS/VS2, MVS COMPLETE, ROSCOE, TSO,
                   INFOMGT,   JES2/NJE,   SCS63   R4.,   CLSS1/R1.,
                   CODEL/R58., DLC6/R1., EP4/R1., NSS1/R1., MAF3/R2.,
                   CNS2/R1., ACF/NCP/R1.2.1., SRM1/R2.

                   Hardware: AMDAHL 470/V7, IBM 4341, COMTEN 3690,
                   IBM/(3271, 3272, 3274, 3275, 3276, 3277, 3278, 3286, 3287,
                   PC'S), TELEX/(274, 276, 178, 284, 286, 287, 289), ATT/(4540
                   DATASPEED   SERIES,   MODEMS/(DATAPHONE   II,
                   RACAL MILGO COMLINK III, CODEX 8200), HALCYON
                   801A, QUESTRONICS 400.

                   Planning Research Corporation
                   Executive Office of the President
                   Washington, D.C.

Feb 85 to Present ***Communications Technician***. Serving as a Communica-
                   tions Technician providing hardware support of government
                   owned equipment such as data terminals, printers, modems,
                   personal computers, and word processors. Telecom-
                   munications requirements include support of data lines,
                   protocol converters, asynchronous switching systems, and
                   front end processors.

Planning Research Corporation (continued)
Executive Office of the President
Washington, D.C.

Software: OS/VS2, MVS, JES2, SNA, VTAM, TCAM, BCAM, TSPF/SPF, M204, SCS63/R4., SLSS1/R1., EP4/R2., ACF/NCP/R1.2.1.

Hardware: Mainframes: Twin IBM 3083's, IBM 4341, DEC PDP-11, DEC VAX 11/70, PRIME 550, DG MV8000.

Front End Units: Twin COMTEN 3650's, IBM 3705, GANDALF PACX IV.

Controllers and Multiplexers: IBM 3274 (41D & 51C), FIBRONICS FM1632, XENER BOX (Models 1 & 2).

Word Processors: IBM Displaywriter System, Decmate, Xerox 860, Lexitron VT1303, Lanier.

Terminals: DEC VT100 and 200, FALCO TS-1, Hazeltine 1510, IBM 3278 and 3279, Honeywell VIP78, CIT 101 and 102.

Modems: Hayes Smart Modem 1200, RACAL VADIC VA212, Bell 212A and 201C, Penril PSH 24 and 2127, GANDALF LDS-120 and 309.

EDUCATION

OS Systems/360 Operators School
USMC Computer Sciences School
Quantico, Virginia

Introduction to Data Communications
CDC, Washington, D.C.

System Fundamentals
NCR/COMTEN, Rockwell, Maryland

One year, Business Major
Northern Virginia Community College

SECURITY CLEARANCE

Final SECRET with USMC
Currently undergoing BI for TS for PRC job

GEORGE LEWIS GRANT
929 Tulip Avenue
Newport Beach, California 92662
(714) 555-1056

OBJECTIVE

To obtain a full-time position in computer graphics or animation using my abilities to communicate with both technical and non-technical individuals.

EDUCATION

MASTER OF SCIENCE degree in **TELECOMMUNICATIONS**, Department of Electrical and Computing Engineering, University of Colorado, Boulder 1985.

Thesis: Interactive Imaging Systems. Described the new applications of technology and social implications of these applications. The primary focus is directed towards applications since industry is without information regarding how interactive imaging systems might be used in the study of art history and in the studio arts. In addition, the commercial arts and business graphics are addressed. Finally, this thesis presents an original traffic study for a campus electronic imaging system.

MASTER OF ARTS degree in **ART HISTORY** and minor in **CLASSICAL GREEK LANGUAGE**, Departments of Fine Arts and Classics, University of Colorado, Boulder 1984.

Thesis: The Leagros Group. Confirmed, with noted exceptions, the organization of the Leagros Group (circa 500 B.C.) as devised by J.D. Beazley; isolated two opposing conceptual trends within the Leagros Group—the crowded and uncrowded panels; and questioned the traditional dating of the Leagran period.

BACHELOR OF ARTS degree in **FINE ARTS,** magna cum laude, University of Colorado, Boulder 1973.

Thesis: The Influences of the Writings of Novalis upon the Painting of Paul Klee.

PROFESSIONAL AFFILIATIONS

Institute of Electrical and Electronics Engineers (IEEE).

EXPERIENCE

AT&T-Information Systems, Video Switching Development Group, Denver. Senior Technical Associate, Graduate Internship June through September 1985.

Set up a dBase III filing system for storage of client information and meetings; computerized a slide system; completed an introduction "C" language; learned the basics of UNIX-CAD; used Edix and Wordix editing systems; wrote technical reports; traveled to the University of Pittsburgh to interview faculty and campus representative for the "Campus of the Future" project; worked on human interface problems on the video switch; and did library research on competitive information.

Department of Fine Arts, University of Colorado, Boulder, Teaching Assistant 1984.

Researched and wrote reports on these specialty areas: Ancient Greek and Roman Art, Modern European Art, American Art, and Art of the Last Decade; taught two weekly classes of survey art history; made up tests and paper assignments.

Office of Veterans' Affairs, University of Colorado, Boulder. Counselor and Administrative Assistant, 1977–1981.

Counseled Vietnam combat veterans for Delayed Stress Syndrome; wrote counseling evaluations concerning veterans at the request of the Veterans Administration; facilitator for the Boulder County Delayed Stress Group; interpreted federal and state regulations on higher education; trained and supervised peer counselors; and wrote news releases for radio and TV.

Accepted to the University of California at Los Angeles to the PhD program in Art History for the fall 1984 term; and accepted to the University of Irvine to the PhD program in Classics to read Greek for the fall 1984 term.

Practicing artist.

REFERENCES AND WORK SAMPLES UPON REQUEST.

FREDERICK J. FREEMAN
596 Marystown Street
Romona, CA 92000
(619) 555-5678

EXPERIENCE

AVCO FINANCIAL CORPORATION • Newport Beach, CA

June 1980 to
Present

Programmer Analyst: Responsible for the modification and tailoring of a generalized General Ledger and Financial System to meet company requirements and standardization. Assignment included analysis, COBOL programming, testing, JCL and modification and documentation. Implemented a new version Bond and Stock System and rewrote customer and operations documentation. Strong user interface. Experience includes IBM 3033 MVS, COBOL, TSO, and JCL. Exposure to VSAM and CICS.

RATNER CLOTHING CORPORATION • San Diego, CA

April 1979 to
February 1980

Contract Programmer: Converted major system programs from OS to DOS. Designed, analyzed, and wrote programs to convert personnel files from various input formats to company standard formats for processing of business applications. Experienced in program, operations, and customer documentation. Conducted maintenance COBOL programming for Personnel and Financial systems. Heavy user interface. Environment IBM 370/135 DOS and DYL-260. Light use of ALC.

MIRA COSTA COLLEGE and PART-TIME WORK • Oceanside, CA

June 1977 to
March 1979

Student/Assistant Manager: Business Administration major.

UNITED STATES MARINE CORPS • San Diego, CA

August 1974 to
April 1977

Programming Manager: Responsible for Personnel, Financial and Logistic System and the supervision of up to 11 programmers. Designed, analyzed and programmed new systems utilizing ANS/COBOL and structured programming. Maintained and modified existing systems. Responsible to direct customer interface. Environment IBM 360/50 OS/HASP with COBOL. Light use of ALC.

June 1967 to
July 1974

Programmer Analyst: COBOL programmer for the maintenance and enhancement of the Marine Corps Personnel and Administrative Systems. Designed and wrote new systems utilizing OS COBOL. Experienced in IBM 370/65 OS/HASP and NCR 304 with COBOL, ALC and MARK-IV.

EDUCATION

University of Washington 1953–1955
Mira Costa College 1977–1978

TRAINING

Accounting, Budget and Cost Accounting, 1980
IDMS, 1980
Advanced COBOL and ALC, 1975
Computer Science School, Quantico, VA, 1972
NCR Programming, 1967

Computer Programmer (Entry Level)

JANE BLACKINGTON
248 North Tigres Loop, #94
Irvine, CA 92714
(714) 555-6574

OBJECTIVE

Entry level programmer with opportunity for advancement to management level.

EDUCATION

Electronic Computer Programming Institute, Kansas City, Missouri,
  graduated August 1985.

Studied fundamentals of Data Processing.

Structured Program Design and Documentation.

Programming in Basic, IBM Assembler, COBOL, CICS, RPG II, OS/JCL, DOS.

Hardware IBM PC, TI 990, IBM 3277, IBM 4381.

WORK EXPERIENCE

Sept. 77–May 81   Long John Silvers

Began as cashier and was promoted through several job levels to Assistant Manager within one and one-half years.

Duties Included:
- Calculated and determined daily food usage and allowances and preparing morning and evening cash.
- Calculated daily and weekly food usage/waste reports.
- Calculated on a monthly basis entire store inventory usage/waste report.
- Calculated and prepared weekly reorder.
- Supervised up to seven employees on shift.
- Interviewed and hired employees.

May 81–Nov. 81   Dobbs House Incorporated
Caterer to airlines at Kansas City International Airport.

Duties Included:
- Prepared morning flight meals for first class and tourist class sections.
- Set up and loaded buffet cabinets to be put on airplanes.

CHARLENE FEMLEY
977 W. Valley View Road
Buena Park, CA 92555
(714) 555-2435

OBJECTIVE	To build a career as a computer programmer/analyst
EDUCATION	*CONTROL DATA INSTITUTE* • Anaheim, CA

Graduate: October 1987 with a 94% GPA
Major: Computer Programming/Operations
Earned an A.S. degree in Data Processing while working part-time.
Completed over 650 hours of intensive training ahead of schedule.

ANTIOCH UNIVERSITY • Santa Barbara, CA
Graduate: June 1979
Major: Women's Studies with emphasis on counseling skills
Earned a B.A. degree while working full-time.

**HARDWARE/
SOFTWARE**

IBM 3083, Zenith PC, CDC 110 Microcomputer
OS/VS JCL, COBOL, CP/M, MS/DOS, TSO, SDSF, ISPF/PDF
Lotus 1-2-3, dBase III, MultiMate
Familiar with BASIC, FORTRAN, 'C', NATURAL, ADABAS SQL

EXPERIENCE

PROGRAMMER TRAINEE. Updated existing programs based on user specifications; analyzed, designed, programmed, documented, debugged, and implemented batch programs using JCL, COBOL, and NATURAL in an IBM mainframe environment.

ADMINISTRATIVE. Developed, maintained and modified filing systems; organized, implemented and administered annual training program based on Nuclear Regulatory Commission regulations for over 100 security personnel; scheduled and coordinated shift changeovers; assisted in recruitment and orientation of new employees; conducted training classes in CPR and security procedures; supervised clerical and security personnel; investigated security and safety violations; computed and prepared payroll and statistical reports.

HUMAN RELATIONS. Established cooperative working relationships with co-workers and upper management; liaison between employees and outside agencies; counseled individuals during crisis and non-crisis situations; listened to grievances and determined proper course of action; analyzed life-threatening situations and implemented appropriate action.

CHRONOLOGY

1988	Insurance Office of America, Inc. • El Toro, CA
1981–86	Diablo Canyon Nuclear Power Plan • Avila Beach, CA
1979	Wells Fargo Guard Service • Santa Barbara, CA
1978	YMCA • Santa Barbara, CA
1975–78	County of Santa Barbara • Santa Barbara, CA
1973–74	City of Glendale • Glendale, AZ

MARYANNE WEISSKOPF
8823 W. Journeyman Circle
Cerritos, CA 90290
(714) 555-4587

SUMMARY RESUME

OBJECTIVE	COMPUTER PROGRAMMER

COMPUTER EXPERIENCE

HARDWARE/UTILITIES. IBM 3090, MVX/XA, PANVALET, INFO/MGMT, UCC1,UCC7, TSO/ISPF, OS/VSE, DOS/VS/VSE, PDF, AND PF.

LANGUAGES. COBOL: structured programming, multi-dimensional tables; VSAM; search, search all, sort, master file update, purchase orders; BAL: sorting, editing, table creation, table search; control breaks; mailing labels; bit/byte manipulation. Adabase; Natural 2; Easytrieve Plus.

CAREER SUMMARY

COMPUTER PROGRAMMER. First American Real Estate Tax Service, Brea, CA. Developed report listings for various county assessors nationwide including Hawaii. Used CAP Frames to generate COBOL code to sort and compare incoming data through the use of VSAM as well as sequential files.

COMPUTER PROGRAMMER. Farmers Insurance Group, Inc., Los Angeles, CA. Worked on Texas Care projects using Assembler to implement changes in updating the mainfile as well as using Easytrieve Plus to initiate the indexing of the mainfile for easier detection of various fields. Implemented change code changes, table changes, and routine maintenance of the CARE Mainfile.

COMPUTER PROGRAMMER. Insurance Office of America, Inc. El Toro, CA. Developed, tested, and implemented tracking reports for various bank auto loans using COBOL, Syncsort, Adabase, and Natural 2. Used IBM 3083 System with TSO/ISPF, PDF, AND PF.

EDUCATION

B.S. Degree, Business Administration. Hawaii Pacific College, Honolulu, Hawaii. Emphasis on Management and Marketing.

Computer Programming. Computer Learning Center, Anaheim, CA (Graduated Cum Laude)

References available upon request.

Computer Systems Analyst

RICHARD L. SCHUETTE   P.O. Box 18155
(714) 752-6001   Irvine, California 92713

PROFESSIONAL EXPERIENCE

SAUDI BASIC INDUSTRIES CORPORATION (SABIC), Riyadh, Saudi Arabia
Computer Systems Analyst 1989–1991

Performed analysis, design, programming, documentation, implementation, and maintenance for batch and on-line systems: payroll, personnel, industrial production analysis, inventory, sales, manpower, and comparison of actual financial performance to plan using COBOL, CICS/VS, and VSAM in an IBM 4361 environment operating VM/CMS and DOS/VSE. Trained in-house staff in structured COBOL and VSAM processing techniques. Maintained CICS systems tables.

FIRST INTERSTATE BANK, Los Angeles, California
Programmer/Analyst Specialist 1988–1989

Senior developer in a team that produced a pilot demonstration of MODEL 204 DBMS by creating an interactive system that delivered periodic and ad hoc reporting to track cost and profitability of commercial banking products. Designed systems model and created data dictionary to establish company standards for new database architecture in an IBM 3090 MVS/ESA environment.

ROYAL SAUDI NAVAL FORCES, Jubail, Saudi Arabia
Senior Programmer/Analyst 1986–1987

Performed analysis, design, and programming, and trained application users in a large military supply system. Created on-line interactive forecasting and shipyard workload tracking systems for marine engineering groups. The network domain included IBM 3083 and 3090 processors using MVS/XA, COBOL, IMS DB/DC, CICS/VS and VSAM.

CALIFORNIA FEDERAL BANK, Rosemead, California
Senior Systems Analyst 1985–1986

Leader of a consumer lending systems analysis group that designed, produced and implemented a reporting system to import loan performance data from a UNISYS source and convert it to a VSAM database in an IBM 3083 MVS/XA environment utilizing COBOL, CICS/VS and DYL280. Trained group staff in MVS/JCL and DYL280.

AUTOMOBILE CLUB OF SOUTHERN CALIFORNIA, Costa Mesa, California
Senior Systems Analyst 1983–1984

Researched, designed and documented system specifications for the rewrite of a large membership system using structured techniques of SDM70 planning and development methodology. Defined conversion of sequential master file to a VSAM database enabling system upgrade with CICS enhancements in an IBM 3081 MVS/XA environment.

FLUOR DANIEL ENGINEERS AND CONSTRUCTORS, Irvine, California
Senior Systems Analyst 1981–1983

Responsible for maintenance and production problem resolution of payroll, personnel and corporate administrative systems. Consulted at construction job sites in remote worldwide locations. Participated in conversion of files, programs and production control procedures from CDC to IBM mainframe systems.

HARDWARE/SOFTWARE

Mainframes

IBM 30XX models using MVS/XA/ESA, JCL, TSO/ISPF, and CLISTS.

IBM 43XX models using VM/CMS, DOS/VSE, JCL, and POWER JECL.

Interface with CDC and UNISYS during conversions to IBM systems.

Application Software

COBOL, CICS/VS and DYL280 utilizing VSAM file structures and IDCAMS utilities.

Databases

IMS DB/DC with DL/1. MODEL 204 DBMS with its User Language.

EDUCATION

University of Washington, Seattle, Washington
  Engineering

Woodbury University, Los Angeles, California
  Accounting and Business Administration

IBM Education Centers, various USA regional Centers
  Wide spectrum of business and technical systems training throughout 13 years
  of employment at IBM Corporation.

GEORGIANNA SMITH SWARTZMEISTER
92 French Creek
Three Arch Cove, CA 92544
(714) 555-2968

PROFESSIONAL OBJECTIVE

To obtain a position utilizing the technical, organizational, and interpersonal skills gained during my past employment as a Contract Administrator.

SKILLS AND QUALIFICATIONS

My experience and education are primarily in the areas of Sales and Marketing, contract administration. I pride myself in being highly motivated, well organized, accurate, and ready to accept challenges. I feel my excellent communication and creative problem solving skills, in addition to my ability to work independently will be an asset to any organization. My technical knowledge includes the following operating systems and software programs: DOS, Unix, VMS, Wordstar, Wordmarc, JET, AIDS, Lotus 123 and DBASE III. Additionally, I am proficient on many different computer systems and peripherals.

EMPLOYMENT HISTORY

1988–Present   EECO COMPUTER, INCORPORATED • Santa Ana, CA
A hotel property management systems firm.

MARKETING COORDINATOR. Liaison with outside marketing agencies, public relations contracts and international distributors. Coordinate trade shows, advertising, brochures, press releases, prepare marketing budgets, track hardware companies' pro-point programs. Created and administered new procedures for documentation inventory.

CONTRACT ADMINISTRATOR. Reviewed all sales and maintenance agreements for configuration accuracy and content. Generated actual sales revenue analysis. Successfully recovered over $500K in bad debt sales revenue. Acted as ECI Users Group liaison, newsletter editor, and leasing company and bank financing coordinator. Processed sales/purchase orders and customer invoicing. Verified customer credit and collected accounts receivables.

1986–1988   AUTOMATED PROCESSING & DEVELOPMENT CORPORATION • Santa Ana, CA
An escrow/title company computer hardware and software systems firm.

CONTRACT ADMINISTRATOR. Complete responsibility for all sales contracts, which included production of installation reports and verification of leasing arrangements. Maintained company database including production of customer reports and mailing labels. Provided customer support for 1099 Real Estate Reporting document processing.

PRODUCT SUPPORT REPRESENTATIVE. Client services, which included hotline support of all products, new and existing document definition, and trouble report follow-up. Interfaced with other departments to assist clients with technical problems encountered while operating various systems. Traveled to new customer sites to present initial visit materials, and analysis of office areas which included requirements for computer space and cabling, and power availability. Additional duties included quality assurance, training of new staff members. Tested and loaded all required software for the installation and conversion specified on contract configuration.

1983–1986   ALBERT C. MARTIN AND ASSOCIATES, Los Angeles, CA
An architectural/engineering/design firm.

ELECTRICAL ENGINEERING DRAFTSPERSON/COMPUTER TECHNICIAN. Responsible for data entry of electrical lighting, footcandle and Title 24 building compliance calculations. Computer aided drafting/graphic input of electrical schematics.

ENGINEERING DEPARTMENT SECRETARY/RECEPTIONIST. Duties included word processing of all departmental correspondence and contracts, maintenance of all files, telephone screening, meeting designation, travel preparations and special project presentation arrangements for a group of 45 engineers. Responsibilities included reception of all company visitors, coordination of conference areas, and directing incoming telephone calls to over 200 extensions.

EDUCATION

HAMILTON HIGH SCHOOL, Los Angeles, CA
(1982 Graduate) General Education

CALIFORNIA STATE UNIVERSITY, Los Angeles, CA
Psychology & Basic Computers

GOLDENWEST COLLEGE, Huntington Beach, CA
Business & Drafting

ORANGE COUNTY SCHOOL OF ESCROW, Yorba Linda, CA
Beginning Escrow Course

NCR Corporation, Dayton, OH
Wordmarc Training

REFERENCES

Professional and personal references are available upon request.

DIANA WILLIAMS PRINCE
899 Riverview Terrace • Brea, California 92500
(714) 555-4875

EDUCATION

9/81–12/87	California State University, Fullerton BACHELOR OF SCIENCE, COMPUTER SCIENCE: 6/87

LANGUAGES/SOFTWARE

Languages	Ada, Assembly, Basic, Cobol, C, Lisp, dBASE II, dBASE III, Fortran IV, Modula-2, Pascal
Hardware	CDC-70, DEC, PDP-11, IBM PC, TERAK, IBM 3081D, IBM 3081K, IBM 3033U, Apple II PC, Toshiba Laptops
Operating Systems	RSTS, NOS, Unix, CICS, IDMS, SAS, Ingres, DOS

EXPERIENCE   INDEPENDENT CONTRACTOR/CONSULTANT

2/88–Present   ***Toshiba America, Incorporated***, Irvine, California
Duties: Write program to test the Toshiba laptops and competitors. Programs were to be written in C and Assembly. Write appraisal reports. Write and review manuals. Revise and update Assembly or C code to function in current industry. Design and develop new features for laptops.

5/87–1/88   ***Quality Telecommunications***, Anaheim, California
Duties: Computerized the office's manual bookkeeping tasks. Installed software to perform in-house payroll. Designed and installed a billing package including accounts receivable functions.

9/86–5/87   ***Austin Foust and Associates***, Santa Ana, California
Duties: Wrote programs to evaluate traffic count data. Programs were to be written in Ada, Basic, Fortran or Dbase III depending upon application. Generated reports after completion of runs.

11/85–12/86   ***WesCorp Federal Credit Union***, Pomona, California
Duties: Wrote programs to analyze Credit Union's account data and format into machine level usage. Included data entry performances.

1/84–11/85   ***Fluor Engineers Incorporated***, Irvine, California
Duties: Evaluated the operating system and generated weekly and monthly reports based on findings. Installed system support libraries. Unified data centers to corporate level. General maintenance of the system. Wrote programs to test and perform the analysis and installation of functions.

PATRICIA BOLSESTA
18699 ARROYO DE GATOS
McLEAN, VIRGINIA 22022
(703) 555-2468

OBJECTIVE   A challenging data processing position in end user or customer support with opportunities leading to management.

HARDWARE/SOFTWARE EXPERIENCE

APPLE Macintosh; IBM PC-XT, PC-AT and compatibles to include: WANG, ZENITH, AT & T. COMPAQ, HP 110, IBM 5520 Administrative System; HP 3000.

PC-DOS, LOTUS 123, SUPERCALC 3, MULTIMATE, ENERGRAPHICS, CHART-MASTER, SIGN-MASTER, DIAGRAM-MASTER, DBASE III PLUS, KNOWLEDGEMAN, REMOTE, and CROSSTALK. APPLE software includes: MACWRITE, MACDRAW, MACPAINT, and MAC-PROJECT.

EMPLOYMENT

ASSISTANT ENGINEER (March 1986–November 1987)
McDonnell Douglass Astronautics Company, 1235 Jefferson Davis Highway, Arlington, VA

Served as the Washington-based liaison for Computer Based Operations within the Tomahawk Weapon System-System Engineering and Integration Agent (TWS-SEIA). Managed the operation and maintenance of the computer facility. Analyzed and recommended the acquisition of all computer hardware and software within the office. Coordinated and integrated Configuration Management requirements with all program disciplines (Engineering, Contracts, Operations, and Customers). Prepared and provided engineering input to change proposals. Collected and verified engineering data into the Configuration Status Accounting System (CSAS). Served on the National Committee to evaluate CSAS system improvements. Represented McDonnell Douglas at Configuration Change Board meetings held at the Navy Cruise Missile Project Office.

MICROCOMPUTER SYSTEMS SPECIALIST (July 1984–March 1986)
Marine Corps Central Design and Programming Activity (MCCDPA)
Small Systems Branch, Quantico, VA

Served as analyst in the design and implementation of microcomputer requirements. Provided consultation in the analysis of systems requirements to end users. Served as liaison with vendor representatives. Coordinated and conducted feasibility studies to determine the practicality of requested microcomputer requirements. Developed cost analysis configurations. Developed personnel training programs.

SYSTEM ADMINISTRATOR (December 1983–July 1984)
Marine Corps Central Design and Programming Activity (MCCDPA), Quantico, VA

Served as the ADP Staff Assistant to the Deputy Director. Provided technical information and training on special projects. Prepared overall project recommendations from a data processing standpoint. Implemented an Office Automation System utilizing the IBM 5520 system. Coordinated the procurement and scheduling of equipment shipments. Developed and implemented a Management Information System. Provided complete documentation and user training on all projects.

EDUCATION   Bachelor of Science: Criminal Justice and Sociology; Psychology Minor
Radford University, Radford, Virginia

Training in Computer Science includes: Data Communications, Database Management Systems, Local Area Networks, System Design and Analysis, and Graphics Applications for Microcomputers.

CLEARANCE   SECRET

BART D. DUNFIELD
527 Ernestine Court
Orange, CA 92800
(714) 555-9233

CAREER OBJECTIVE To obtain a programming position in which I may utilize my technical expertise in engineering with my data processing background.

EDUCATION B.S. in Electrical Engineering, University of Kansas, 1970. Includes a course in Fortran programming.

Computer Learning Center, 9/84–3/85.
Graduated in Programming, G.P.A. 92%
Extensive training in application programming in a DOS/VSE, OS/VS1 environment on an IBM 4341 computer. Languages: COBOL RPG II BAL

The assigned problems covered a variety of applications including updating inventory master files, generating error reports and internal sorting and processing of records. Completed the optional Operations Seminar.

MILITARY DUTY Honorable discharge from U.S. Army. Security clearance, Secret. Last station Tongduchon, Korea.

EMPLOYMENT
1982–1984 INDUSTRIAL WASTES INTERNATIONAL • LEAWOOD, KS
Position: *Area Representative*

1979–1982 FLUOR CORPORATION • IRVINE, CA
Position: *Senior Electrical Engineer*

1975–1979 OCCIDENTAL ENGINEERING COMPANY • IRVINE, CA
Position: *Senior Electrical Engineer.* Occasionally designed, wrote and debugged engineering application programs in Basic.

1971–1975 GAS SERVICE COMPANY • KANSAS CITY, MO.
Position: *Senior Project Engineer* for multi-engineering projects. Occasionally designed, wrote and debugged engineering applications in Fortran.

REFERENCES Will be furnished upon request.

WALTER R. RODRIGUES

20456 Ranch Catalina
Laguna Hills, CA 92650
(714) 555-2533

EDUCATION

California State Polytechnic University, Pomona
Bachelor of Science in Electronics Engineering, 1976

WORK EXPERIENCE

Altos Computer Systems • La Palma, CA   Sept 1990 to Apr 1991

District Sales Manager. Responsible for recruiting and managing Value Added Resellers (VARs) to sell UNIX based multiuser computer systems in Southern California and Arizona territory. Signed 8 new resellers in first 6 months and was
100% of quota in March.

Tektronix • Irvine, CA   Feb 1988 to Jun 1990

Sales Engineer. Responsible for recruiting and managing Value Added Resellers
(VARs) for graphics workstations, terminals, and printers. Signed 8 new VARs in
the first year in virgin Southern California territory. Developed and signed $2.5M
strategic account. Achieved 110% of target in fiscal year ending May 26, 1990.

Metheus Corporation • Costa Mesa, CA   Sept 1983 to Feb 1988

Senior Sales Representative (8/85 to 2/88). Developed Southern California sales
territory for OEM graphics display controllers in the commercial and government
sectors. Also managed Manufacturer's Reps in the remainder of the Southwest territory. Achieved 110% of quota and attended President's Club in first year. Was
105% of quota with major new OEM's in last year.

Technical Support Specialist (9/83 to 8/85). Pre-sales and Post-sales support for
Western U.S. Responsible for demos, presentations, installations and training.
Signed first new OEM account while in this position which resulted in promotion
to Sales.

Data General Corporation • Santa Ana, CA   Jan 1981 to Sept 1983

Systems Engineer. (Secret Clearance) Responsible for sales support including demos, presentations, benchmarks, configurations, and other pre-sales activities, as
well as post-sales consulting, training, and problem solving.

Interstate Electronoics Corporation • Anaheim, CA   Apr 1976 to Jan 1981

Electronics Engineer. (Secret Clearance) (1/78 to 1/81) Responsible for design and
support of various FORTRAN programs for missile tracking simulations and system tests. (4/76 to 1/78) Coordinated equipment interfaces, both mechanical and
electrical, between Interstate and other contractors on the FBM program.

Charles J. Jameson
968 E. Boulevard Way
Buena Park, CA 92357
(714) 555-9812

GOAL   Environmental/Occupational Health and Safety.

GENERAL SUMMARY

M.S. Environmental/Occupational Health Science. Demonstrated experience in industrial hygiene, hazardous exposures, fire safety and loss prevention planning and controls, product safety, training, communications, research, recommendations, and written reports.

BUSINESS EXPERIENCE

1990 to Date   CNA Insurance Companies • Brea, CA.
Loss Control. Responsible for conducting surveys of businesses, buildings and construction sites to gather information on hazards and controls related to insurance coverage provided. Interfaced with owners, tenants, agents and underwriters. Prepared written reports and recommendations.

1988 to 1990   AETNA Casualty and Surety Co. • Glendale, CA.
Engineering Representative. Responsible for conducting surveys of buildings, businesses, and construction sites to gather information on hazards and controls related to insurance coverage provided. Interfaced with owners, tenants, agents, and underwriters. Prepared written reports.

1986 to 1987   Moorpark College • Moorpark, CA.
Tutored learning-disabled students in math, science, and English. Received a certificate of appreciation as a classroom aide.

Concurrent: Reviewed underground tank-leak detection plans for legal compliance, monitored hydrocarbon levels in air and soil, and processed computer data for Ventura County Environmental Health Department.

1985 to 1986   Office of Environmental Health and Safety, Cal State, Northridge.
Assisted the Director in investigations, monitoring, and prevention programs. Operated air monitoring devices, conducted fire prevention surveys, and conducted safety inspections. Compiled data and prepared reports.

EDUCATION   M.S. Environmental/Occupational Health Science. Cal State Northridge. 1987

B.S. Resources/Planning and Interpretation. Humboldt State University

A.A. Liberal Arts. Moorpark College. Moorpark, CA. Dean's List

Aetna Institute: Loss Control Training Program

Commercial Insurance: Loss Control, Industrial Safety & Hygiene.

PERSONAL   Open to relocation. Enjoy outdoors, photography, and reading.

REFERENCES   Available upon request.

Barney T. Roobaugh
136 S. Avenida de Naranja
Brea, CA 92987
(714) 555-1098

GOAL Computer Technology, Hands-on Supervisor.

GENERAL SUMMARY

Ten years of demonstrated technical experience with emphasis on computers and peripherals, including installation, trouble-shooting, maintenance, and repair. Managed field service engineers. Provided customer service and applications information. Ordered and maintained inventory.

BUSINESS EXPERIENCE

1989 to Date   Word Processing Brokerage/LANtek • Costa Mesa, CA.
Field Engineering Manager. Directs and provides instruction and assistance to field engineers; provides customer service information on applications, new products and computer systems maintenance; negotiate, purchase, and maintain inventory on parts, hardware and software.

1986 to 1989   OASYS • Des Moines, IA.
Field Engineering Manager. Directed and provided instruction and assistance to six field engineers in three states; provided customer service information on applications, new products and computer systems maintenance; negotiated, purchased and maintained inventory on parts, hardware and software.

1985 to 1986   Norstan • Oklahoma City, OK.
National Senior Field Engineer. Responsible for handling critical accounts throughout the Midwest and South; troubleshooting hardware and software applications; evaluating changes; recommending solutions, and coordinating with customers.

1982 to 1985   Spaulding • Boston, MA.
Field Engineer. Responsible for installation and repair of computers and peripherals. Involved in customer service, necessary paperwork, ordering and maintaining inventory.

Prior: Installed and repaired computers, peripherals, copiers and microfilm equipment including parts inventory and customer service.

EDUCATION   A.A.S. Electronics Technology. Niagara County Community College. NY. Continued Studies and Hardware Training: Telecopiers; Checkwriting equipment; Word-processing systems; Electromechanical scales; copiers and microfilm; sheetfeeders; NBI printers; Qume; Diablo; Printronix; Centronix; Fujitsu; Epson; Xerox, H.P., Canon and Apple Laser printers; LANs; ICUs; NBI UNIX operating systems; office automation systems; NBI field engineering software; minicomputer and UNIX repair; Kurtzwiel, DEST and P.C. Scan Optical Character Readers; Rutishauser, BDT, Ziad, Qume and Datamarc Sheet Feeders; Mini- and Micro-Computers.

PERSONAL   Enjoy fishing, archery, golf, and computers.
SNA: Service Network Association.
AFSM: American Field Service Managers.

FRANKLIN P. EBERHARDT
20383 Barnsley Square
Hawaiian Gardens, CA 90484
(213) 555-9696

SENIOR FINANCIAL MANAGEMENT

■ Eighteen years progressively responsible experience attained with established leader in Financial Services Industry. Particular strengths include Financial Modeling and Planning, Cash and Asset Management, Financial Reporting, Budgeting and Forecasting, EDP Auditing and Risk Management.

■ Demonstrated effectiveness developing data processing systems for Payroll, AP/AR, Inventory and Cost Control, Capital Depreciation and Amortization, and General Accounting Applications.

■ Solid success increasing productivity, assuring quality, and building efficient teams while enhancing bottom line results.

CAREER HIGHLIGHTS

OWNER/CONSULTANT ■ D & B Computer Services

Design customized automated financial models and systems for clients in restaurant, legal, healthcare, retail, service and manufacturing industries. Began part-time in 1984, expanding to full time in 1989.

■ Implemented Payroll, AP/AR and Inventory Systems enabling clients to maximize return on capital, lower interest expense, and reduce costs by streamlining operations.

■ Created "in-house" accounting function for restaurant client, substantially reducing external accounting firm fees.

■ Increased operating efficiency for food services client by providing hourly breakdown of sales, making employee scheduling more economical.

VICE PRESIDENT ■ Merrill Lynch & Co.   1971–1989

Rose rapidly through variety of assignments to positions of increased responsibility within international financial services firm.

■ Developed Cash Management System to clear funds nationwide ensuring daily efficient use of corporate cash.

■ Managed $20 million in corporate investments including Retirement, Profit Sharing and Pension Fund Accounts for Fortune 700 clientele.

■ Prepared and controlled Sales and Operational Budget for Southern California Region (75 offices) totalling $850MM.

■ Managed investments totalling $50MM consisting of CDs, Bonds, Partnerships, Real Estate, Venture Capital, Equities, Insurance and Annuities.

■ Administered branch offices consisting of 63 Financial Consultants and 22 Accountants improving efficiency rankings from bottom to top 10% within 6 months.

■ Directed technical and accounting staff of 150 as "CFO" of new subsidiary, processing 2,000 investment transactions daily and generating monthly profits of $400,000.

■ Implemented controls accurately tracking all client assets through processing system.

■ Performed EDP audits at 300 locations nationwide ensuring data system integrity and developing procedures improving operational efficiency.

EDUCATION

BS, Economics, BS, Finance, John Carroll University.

PERSONAL

Enjoy tennis, skiing, softball, chess, community volunteer work.
Professional references available upon establishment of mutual interest.

MICHAEL S. AXFORD
45022 Sentinal Avenue
Laguna Niguel, CA 92977
(714) 555-2853

EXPERIENCE   BURLINGTON NORTHERN AIR FREIGHT ■ Irvine, CA

3/84 to 3/86   Information Center Manager/Systems Analyst. Supervised the formation of an Information Center. Responsible for all areas of user support, including training, coordinating software and hardware requirements, evaluation and cost analysis of "end user" software and microcomputer peripherals, etc. Developed several ad hoc marketing programs for senior management using FOCUS and SAS.

FLUOR CORPORATION ■ Irvine, CA

4/80 to 3/84   Supervisor/Systems Analyst. As Coordinator and Technical Consultant was responsible for user training and technical support in software languages which included: SAS, IFPS (Interactive Financial Planning System), AUTOTAB, CULPRIT, and ANSWER/DB. User consultant for applications involving TSO/ISPF, Dialogue Manager, CLISTs, and JCL. In the areas of design and development was responsible for the specification and testing of several Management Operating Report Programs. Designed and developed COBOL programs for a Cost and Budget/Financial Forecasting System.

EDUCATION   Bachelor of Science, Pomona College, Claremont, CA, 1979
Major: Mathematics, Minor: Psychology

Orange Coast/Saddleback Community Colleges, 1981–82
Studies included courses in Computer Science, Accounting, and Finance

HONORS/AWARDS   Dean's Honor List, Pomona College, Claremont, CA
Candidate for Senior Thesis Honors

BRITT G. REID
326 East Boulevard Drive
Glenview, Illinois 60099
Home: (312) 555-3749
Office: (312) 555-1029

CAREER SUMMARY

Fifteen years of increasingly more responsible positions in Information Systems Management with Fortune 100 companies, involving department management, international planning, budgeting, recruiting, operations, and system development and support.

BUSINESS EXPERIENCE

KRAFT GENERAL FOODS, Phillip Morris Companies • Glenview, IL   1988 to Present
A $22 billion multinational food products company.

Director, Systems Planning and Architecture—International
Responsible for review, coordination, and approval of strategic system plans of international and independent business units. Also recommend and assist in implementation of MIS solutions.
Key accomplishments:
- Coordinated and approved 1988–1989 strategic MIS plans for 10 international and four independent business units.
- Recommended, and assisted in implementing computerized solution to financial preparation deficiencies in London export office. Resulted in conversion from manual to mechanized processing with more timely, accurate, and complete information.
- Finalized hand-held terminal solution for major depot reorganization project at Raft, Italy, including communications network upgrade.
- Assisted in mainframe upgrade solution for Kraft, United Kingdom, including capacity analysis. Resulted in $40,000 cost savings.
- Participated in developing common systems architecture solution for new systems to be developed and migrated between Kraft Venezuela, Mexico, and Spain. Leveraged development costs between three units.
- Finalized common financial and manufacturing package solution for three divisions of Frozen Food Group including reporting consolidation. Lowered package cost, enhanced installation exchange between divisions and ensured consolidation capacity.
- Analyzed consolidation of Pollio Division's MIS applications into Corporate mainframe, reducing MIS and other division business costs.

WICKES LUMBER DIVISION, Wickes Companies • Vernon Hills, IL   1985 to 1988
A $1 billion retail merchandiser of building products.

Director, Management Infomration Services
Supervised division MIS activities including operations, system development, and support with staff of 75. Member, Wickes MIS Executive Council.
Key accomplishments:
- Phased installation of IBM Series/1 computers in 250 lumber stores (SNA) across 38 states with $25 million budget.
- Prepared division's first strategic MIS plan using Nolan & Norton's techniques.
- Installed Cullinet's IDMS database for merchandising systems, resulting in first operational database implementation.
- Replaced accounts payable system with Data Design's software, centralizing processing of both retail cost and revenue data.
- Completed MIS decentralized plan resulting in stand alone operations.

BAXTER TRAVENOL LABORATORIES • Deerfield, IL   1979 to 1985
A $6 billion international distributor of hospital supplies.

Group Manager, Financial Systems (1983–1985)
Responsible for all Corporate financial computer systems. Supervised $2 million budget and staff of 36.
Key accomplishments:
- Installed Worldwide Financial Consolidation System, resulting in first computerized reporting globally.
- Replaced Accounts Receivable System with CARMS, using Arthur Andersen's Method/1 (System Development Life Cycle Method).
- Restructured MSA's General Ledger System for improved reporting.
- Installed PC-based Budget and Spending System, for improved control.
- Installed McCormack & Dodge's Fixed Asset System and Capital Expenditure Control System.

Manager, EDP Audit (1979–1983)
Responsible for EDP audits of all domestic and international computer installations. Chairman, EDP
Audit Committee of Pharmaceutical Association.

FMC CORPORATION • Chicago, IL   1977 to 1979
A $4 billion multinational equipment manufacturer.

Senior EDP Auditor
Responsible for all EDP audits, domestic and international, including computer operations, production
applications, and development systems.

ARTHUR ANDERSEN & COMPANY • Chicago, IL   1974 to 1977
A multinational audit firm with world's largest consulting practice.

Senior Consultant

EDUCATION

BSBA, Accounting, Roosevelt University, Chicago, IL, 1974
Strategic Systems Planning, Harvard Graduate School of Design, 1986
Telecommunications Network Planning, Arizona State University, 1988
Additional training courses, including: Assembler and COBOL languages, Audit Software, Systems
Design and Installation, Advanced Operating Systems, Arthur Andersen & Company.

MILITARY

U.S. Army, Vietnam, Officer Replacement Division (1968•1969)

OTHER

Chairman, Board of Trustees, St. Augustine College
Secretary, Information Planning Association
Former Member, IBM Business Advisory Committee on Computer Training to the Disabled

THURMAN J. KENNEDY
4805 Lom Lane
National City, CA 92100
(619) 555-0098

EDUCATION   Santa Ana College
Associate of Arts awarded June, 1976
Major: Computer Science

HARDWARE   DEC VAX Family, DEC PDP-11 Family, DEC System 10;
IBM AT; XEROX Sigma-7

SOFTWARE   VAX/VMS; RSX-11M/M PLUS; RSTS/E; MS-DOS

LANGUAGES   SPL LANGUAGE, MACRO-11; FORTRAN F77; PDP-11
BASIC-PLUS-2, VAX-11 BASIC

EXPERIENCE   Data Processing Design, Inc.   May 1985–Present

Applications Programmer

Responsible for design, development and implementation of
touch tone telephone response applications on Vax-11/750
and DECtalk (DTC01). Conducted demos with customers.
VAX-11 BASIC and Datatrieve

Responsible for sysgens under RSX-11M 4.2 /M PLUS 3.0
with DECnet. PDP-11

Assisted in the migration of software products from
VAX/VMS to PDP-11 RSTS/E. SPL Language, MACRO-11

Development of document conversion applications for word
processing documents. IBM AT, VAX-11 BASIC

Oak ADEC, Inc.   Nov. 1982–Nov. 1984

Software Engineer

Responsible for integration and testing of real-time data
acquisition applications for large scale energy management
systems.

Conducted system generation under RSX-11M/M PLUS for
customer sites. Conducted demos and testing of software
applications. Extensive travel to sites.

Responsible for managing all in-house computer systems
and peripherals.

ERNEST DARLINGER

Microcomputer Programming Consultant

P.O. Box 7562   Thousand Oaks, California 91399   (818) 555-8514

OPERATING
SYSTEMS
- MSDOS
- OS/2

PROGRAMMING
LANGUAGES
- C
- Assembler
- MS Windows
- dBase
- Clipper
- BASIC

HARDWARE
- Microcomputers (8086-80486 chip technology)
  Monochrome, CGA, EGA and VGA/SVGA Monitors

SPECIAL
SKILLS
- Application Development
- Database Design
- Software Integration
- Memory Resident Programs
- Mixed Language Programming
- Utilization of 32 bit technology
- Mouse Interfacing
- Graphical User Interfaces
- MSDOS-OS/2 Conversions
- Device Drivers
- Image Processing

PROJECTS

FANTASIA PERSONNEL: Converted *MSDOS* database system to OS/2 network. Revised DOS utility programs to run under *MS Windows*.

GTEL: Programmed and implemented a salesman compensation/reporting system for GTEL's complex hierarchy of territories, personnel, product groups, quotas, incentives, bonuses, etc. The program contained a distributed database interface for mainframe to micro transfer, and system enforced external backup procedures. Used *Clipper* and *Microsoft C* on 3Comm network of 386s.

THE STRATEGIC MARKETING GROUP: Formulated code to update hotel marketing program from manual spreadsheet to fully automated version with interface to external graphics software. Used *Microsoft C*, *Quick C*, and the *Microsoft Macro Assembler* for full family of IBM PCs and monitors.

PEPPERSTONE CORPORATION: Created billable hours tracking/reporting system for legal document production using *Bourne Shell* on UNIX 560 minicomputer network.

Designed attorney timesheet processing system on IBM PC network featuring communications interface to accounting software on off-site UNIX minicomputer system. Used *Advanced Microsoft BASIC*.

Developed research utility for LEXIS-type text retrieval of in-house legal documents using *dBase III+* and *Microsoft C*.

HORVITZ, LEVY & AMERIAN: Created automatic backup program for two minicomputer networks (UNIX 560 and NBI) with interface to coordinate process management between operating systems. Utilized *UNIX Bourne Shell* and *NBI Advanced Stored Keystrokes* languages.

Designed and implemented memory resident utilities for network of IBM compatible PC's using *Microsoft Quick C* and *Microsoft MACRO Assembler*.

Developed tools for software and/or operating systems integration using *Microsoft C*, *dBase III+* and *Microsoft MACRO Assembler*.

SOUTHERN CALIFORNIA RAIL CONSULTANTS: Programmed and implemented tracking systems for correspondence, inventory and schematics involved in the Metro Rail project. Utilized *dBase III+*.

MICHELLE BURTON & ASSOCIATES: Formulated critical components of personnel industry package: user interface, data handling routines. Used *Microsoft C* and *Microsoft MACRO Assembler*.

IBM/SECURITY PACIFIC BANK JOINT PROJECT: Programmed *MSDOS* user interfaces. Trained groups/individuals in *MSDOS* and *Displaywrite* word processing. Administered hotline technical support.

OTHER ACHIEVEMENTS

- Developed comprehensive package of *Clipper* programming tools.
- Devised code module to correct major bug in NBI's footnote renumbering program. *(Received commendation from NBI.)*
- Created user interface programming library using *Microsoft C* and *Microsoft MACRO Assembler*.

DAVID W. O'CONNOR
9235 South Charleston Way
Arcadia, CA 91555
Home: (714) 555-9079 / Office: (714) 555-4967

OBJECTIVE

MIS Management/Operations/Project Management. A responsible management position accountable for the efficient and effective operation of a data processing center or department, or the successful completion of a program or project to enhance productivity and profitability.

SUMMARY

A data processing professional with 22 years of increasingly responsible management experience, with 13 years in data processing management and an advanced education. Extensive hands-on experience in managing computer facilities, planning local area network installations, managing and coordinating data processing projects, developing long-range information systems plans, forecasting and control of $600K operating budgets, and directing technical and clerical staff of 35 people. Capabilities also include preparation of contract proposals, vendor selection and contract oversight.

RELEVANT EXPERIENCE

OPERATIONS MANAGEMENT Managed three computer centers, including an IBM 4381 installation. Responsibilities included computer operations, applications program development, systems programming, local area network administration, office automation, telecommunications, and small computer support. Administered $3.5M of automatic data processing equipment and a $600K annual operating budget.

STRATEGIC PLANNING Developed a long-range organization-wide computer systems plan that effectively coupled computer systems requirements and budget requests with the organization's strategic goals and objectives.

LOCAL AREA NETWORKS Planned, developed the procurement specification, contracted, installed and maintained a modern $2M, 250 PC local area network in a 400-person office. System enhanced productivity by providing users direct access to mainframe databases and allowed the sharing of peripheral devices. The network replaced an antiquated, unreliable minicomputer-based word processing system and saved over $100K annually in maintenance costs.

DATABASE MANAGEMENT SYSTEM CONVERSION Coordinated the conversion of several large databases into a modern relational database management system on an IBM mainframe computer. This was a multi-year project which resulted in the capability to store, access, retrieve, and manipulate information in a much more responsive and timely manner.

MAINFRAME UPGRADES Directed replacement of an IBM 4341 computer with a more advanced IBM 4381. Tasks included creating needs document, purchasing hardware, coordinating replacement schedules, installation and checkout. Managed several disk and memory upgrades.

FACILITY ENHANCEMENTS Oversaw two facility remodelling projects that involved the construction of a new computer room and installation of a new mainframe computer in one case and the enlargement of an existing computer room for a new minicomputer in the other case. Monitored the contractor work in both cases through successful completion.

OPTICAL IMAGING Responsible for an on-going project to acquire a PC-based optical disk image storage and retrieval system that uses state-of-the-art technology for a document processing office. Developed requirements, assembled system specifications, and created the request for approval. System will afford the capability to reduce storage space requirement and improve data retrieval times, while creating paperless file systems.

PRESENTATION GRAPHICS Installed a PC-based briefing system called VideoShow. Worked with the vendor, demonstrated the system to the agency's top-level management and staff, coordinated procurement of the software and hardware, integrated the system components, trained users, and developed maintenance plans. The system has allowed briefers to put together effective, professional-looking briefings in a minimum amount of time.

TRAINING Managed all activities relating to implementation of an in-house training program that included hands-on training on PCs and various software packages. Oversaw training of over 350 personnel and saved an estimated $70K in contract training expenses.

CAREER DEVELOPMENT

CHIEF, OFFICE OF COMPUTER SYSTEMS, Norton Air Force Base, CA
Lieutenant Colonel, United States Air Force   1967 to Present

EDUCATION

M.B.A., Management Information Systems   1973
University of Colorado, Colorado Springs, CO

MASTERS DEGREE, Management   1981
US Naval College of Command and Staff, Newport, RI

BACHELOR OF SCIENCE, Accounting   1966
California State Polytechnic University, Pomona, CA

CERTIFICATE OF DATA PROCESSING (ICCP)   1988

Numerous academic courses to develop professional and personal skills   1970-1985

References Available Upon Request

CALVIN G. SIMONSON
8989 Tigertrail Lane • Fountain Valley, CA 92699
(714) 555-0987

SUMMARY

Fourteen years experience in all phases of Database Management/Programming in Paradox, computer systems and LAN integration, Systems Analysis and Project Management; design, implementation, beta testing, and expansion of Local Area Networks using Novell's Netware and Banyan's Vines.

EXPERIENCE

TRIPLE T INC • Santa Ana, CA   November 1992 to Present

Triple T Inc. subcontracted to DEC who placed my services at Hughes Electro Optical in El Segundo. Member of CAE Network Support group involved in Network downsizing, movement, and restructuring. Duties involved premove inventory of computer configuration and updating software and hardware components; postmove troubleshooting and end user network support and systems reconfiguration. Also involved in response center open call resolution program. Involved in reconfiguration and consolidation of multi server, multi vendor, Novell Network of over 25 servers and 1,000+ workstations. Worked with telecom to troubleshoot segments with traffic overload using sniffer.

SLATER SPECIALTIES • Westminster, CA   March 1977 to November 1992

Computer consultant specializing in Systems Integration, Data Base Management/Programming and On-Site Training; Installation, Integration and Trouble Shooting LANs.

Contract Positions and Projects Worked

DANIEL FREEMAN MEMORIAL HOSPITAL, L.A., CA

Analyzed, designed, programmed (using Paradox) a Medical Staff Roster application, converting from an old DBASE program and an old Dictaphone program.

CHILDRENS HOSPITAL of LOS ANGELES, QA DEPARTMENT • L.A., CA

Analyzed, designed, programmed (using Paradox), a Quality Assurance, Risk Management Data Base consisting of multiple modules (UP, Denial, Infect, Risk, Productivity, Catheter, etc.) running on a Banyan Vines LAN; Installation, Trouble shooting of the Banyan Vines LAN; Systems Administration of Banyan Vines LAN; Training of personnel on LAN and stand alone applications.

ARCHIVE CORPORATION • Costa Mesa, CA

Analysis, troubleshooting and expanded Novell Advanced Netware 286 SFT Local Area Network running on an Arcnet network; reconfigured and installed workstations; layout and setup new workstation locations; pulled cable; specified, purchased, configured, and installed workstations.

MICRO TEMPS • Tustin CA.

Contracted out to So. Cal. Edison, SONGS as systems analyst, micro computer trainer and member of team developing and implementing a 350-node LAN. This network has since expanded to over 1700 workstations and over 100 servers.

CARDINAL PRODUCTS INC. • Santa Ana, CA.

Installed PC-based MRP and accounting systems; set up and taught employees to use word processing, Calc, tracking and accounting software. Designed, wrote and implemented a computerized sales tracking and marketing program. Wrote customer job costing and job quoting program.

PUBLICATIONS

From 1988 to 1990 wrote monthly computer column for the BARNSTORMER and for the SPECTRUM INDEPENDENT (Local Business Newspapers). Wrote documentation and technical manuals for various computer systems and applications.

HARDWARE

IBM PCs and Clones, IBM P/S 2, 386 and 486 systems, COMPAQ, CARDINAL PC, Novell Dedicated File Server, Macintosh Systems, WANG dedicated word processor, XEROX systems, ALPHA Micro system, DEC MicroVAX, DEC Rainbow, AST TURBO LASER, HP Laserjet II, Bubble Jet, Paint Jet and other related printers, plotters, and peripheral equipment.

SOFTWARE

PC/ MS DOS, Novell Netware, BANYAN Vines, WordStar, WordPerfect, Ventura Publishing Software, Page Maker, LOTUS 1-2-3, LOTUS Symphony, Excel, Quatro, dBASE, R:Base, PARADOX, Norton Utilities, MicroSoft Windows, PC Tools, MACE Utilities; CCMAIL, SmartCom, ProCom Plus; AutoCAD; SideKick, ORG, Utilities; PeachTree General Accounting, PeachPak4, PeachTree Complete, CPA+, DAC Easy Accounting, Star Accounting Partner, CYMA Shoebox Accounting, "big" CYMA, SBT Accounting Modules; CopyWrite, CopyII PC; MarketFax, and others.

LANGUAGES

BASIC, BASICA, MS BASIC, GW Basic; LAL, FRED, HAL, Lotus Macro Language, dBASE III, dBASE III Plus, Clipper Compiler, PARADOX APPLICATION LANGUAGE, R:Base Programming Language; BATCH.

EDUCATION

Golden West College • Huntington Beach, CA. (A.A. in Business-1967)
Orange Coast College • Costa Mesa, CA.
Coastline Community College • Huntington Beach, CA.
Long Beach Community College • Long Beach, CA.
Cal State College • Long Beach, CA.
Cal State University • Long Beach, CA
U.C.L.A. Extension • Los Angles, CA
The Cobb Institute • Los Angles, CA (courses in LOTUS 1-2-3 and Symphony)
Microrim Inc. • Redmond, Washington, (courses in programming in R:BASE products)
Novell Inc. • Provo, Utah; (NetWare Authorization Courses)
Coastline Community College • Huntington Beach., CA. Network Certification courses.

ALEXANDER B. WARMATH
907 S. Sunnyside Avenue
Peoria, IL 61600
(309) 555-2944

OBJECTIVE

Design and implementation of automated reasoning programs and expert systems in artificial intelligence applications. This could include the interfacing, design and control of physical devices, operating systems, and databases.

EDUCATION

M.S. Computer Science, Bradley University, Peoria   May 1985
   GPA: 3.42 of 4.0
   Emphasis in artificial intelligence, database, and operating systems.

B.A. Psychology, May 1975, Bradley University, Peoria
   GPA: Overall 3.03 of 4.0. Major 3.25 of 4.0
   Emphasis in industrial psychology and science.

Languages used: Pascal, Prolog, Lisp, FORTRAN, C, BASIC, COBOL, SNOBOL, Compass and Z80 assembler.

Systems used: IBM 360, IBM 370, CDC Cyber171, DEC PDP-11, IBM PC, ITT 3B2, Z80 based microcomputers.

Operating Systems: IBM DOS, CDC NOS, UNIX, PC DOS and CP/M.

EXPERIENCE

Bradley University Computer Science Dept. • Peoria, IL   Spring 84, 85
- Graduate Assistantship: assisting in teaching students CP/M, and Z80 coding and debugging.

Caterpillar Tractor Company • East Peoria, IL   1977–1982
- Completed two-year sheet metal manufacturing apprenticeship in the top 4% of the class.
- Worked as tractor and diesel engine assembler, operated drill, mill, sheet metal forming machines and designed and installed sheet metal products.

K-Mart • Peoria, IL   1973–1977
- Mechanic, also Service Manager on weekends.

HONORS AND ACTIVITIES

Dean's list several times • Commodore Computer Club.

INTERESTS

Microcomputers, automechanics, repairing anything.

AVAILABILITY

June 1, 1985. Interested in relocating.

BERNARD W. WINGLASS
42634 Cabrillo Place • Industry Hills, CA 91399
(818) 555-2945

EXPERIENCE

The Marquardt Company • Van Nuys, CA, *Director of Operations*   1989–1990

$150M high tech government contractor—Major accomplishments include the turn around of a difficult situation concerning a failed Contractor Procurement Survey Review and Method C Program, resolution of negative cash-flow problems with customers/suppliers, and implementation of a staffing and training program resulting in a substantial increase in product quality acceptance and expediting manufacturing.

Unisys Corporation • Camarillo CA, *Director of Contracts*   1986–1989

Developed computer hardware/software air traffic control systems for commercial and government applications—Major accomplishments include organizing, staffing, and directing a newly formed Contracts Department capable of managing a wide range of complex $800M contracts. Formed teaming arrangements, joint ventures, and negotiated transactions with AT&T, GEC-Marconi, Lockheed, Hughes, IBM, Dept. of Defense, Boeing, McDonnell-Douglas.

Ball Corporation • Boulder, CO, *Vice President of Admin.*   1976–1986

Directed the Finance, Contracts, Purchasing, MIS, Legal, Human Resources and Facilities Departments of a $1B, Fortune 500 Corporation. Initially joined the Ball Aerospace Systems Division as Director of Contracts where I established policies, procedures and management controls for the sales of Communications Satellites to NASA and DOD. Received recognition and promotion for performance to Director of Administration (Corp. Staff Position) involving corp. acquisitions, divestitures and foreign operations. In 1983, was promoted to VP-Admin. leading to several assignments. Increased revenues and turned around declining sales and negative profitability of three commercial divisions.

System Development Corporation, *Manager of Contracts and Finance*   1971–1976

EDUCATION

University Of Denver Law School—Juris Doctorate Degree(J.D.)
University Of Colorado—Master Business Administration (M.B.A.)
University Of Denver—Bachelors Degree, Accounting

MEMBERSHIP

Instructor—Business Law, West Coast University, Colorado Bar Assoc., National Contracts Management Association-Past President Colo., APICS, Board of Directors—University of Colorado, Executive MBA

INTERESTS

Golf, tennis, skiing, bridge, music

REFERENCES

Available upon request

LORILEI KILBOURNE
1628 N. Magic Kingdom Way
Anaheim, CA 92710   (714) 555-9837

CAREER OBJECTIVE

A Systems Analyst position involving system configurations, installations and consulting in pre- and post-sales environments.

EDUCATION

California State University, Sacramento
Degree—Bachelor of Science–May, 1983
Major—Business Administration
Concentration—Management Information Science
Certificate in Computer Science—March, 1983

QUALIFICATIONS

Languages/Operating Systems: COBOL, FORTRAN, PASCAL, C-BASIC, M-BASIC, S-BASIC, PL/I, JCL, CP/M

WORK EXPERIENCE

CUSTOMER SUPPORT ANALYST
Wang Laboratories   11/84–present

Solve hardware, software, operating system, system utility, and telecommunication problems customers have with Wang's VS system. Perform demonstrations, figure system configurations, and provide pre-sales support for prospective VS customers.

CLIENT SERVICE REPRESENTATIVE
Automated Data Processing (ADP)   12/83–11/84

Provided software support for clients using ADP's Accounts Payable batch system. Also provided software and limited hardware support for clients using Microdata and IBM equipment for ADP's on-line Accounting Services.

PROGRAMMER/ANALYST
World-Wide Plant Distributors   4/83–6/83

Designed and programmed an accounts receivable and an inventory system using a Kaypro II microprocessor.

PROGRAMMER/ANALYST
Honeywell Comunication Services   06/82–4/83

Procured, designed, and programmed a sales system using a Kaypro II microprocessor.

BARBARA E. WOODBINE
321 East Tanglefoot Drive, Apt. 6D
Anaheim, CA 92800
(714) 555-9866

EDUCATION   BS
Mathematics with minor in Computer Science
University of Illinois, 1974

EXPERIENCE   SAV-ON DRUGS
Anaheim, CA

1/79 to Present   *Programmer/Analyst.* Currently modifying and installing the MSA payroll package as the senior programmer/analyst on a team of two. All programming and development is done in CO-BOL. System environment is an IBM 4341 under OS/VS1 and VM. Know and use DYL-260. Maintaining the current payroll system during the conversion to the MSA package. Present payroll system is being run on an NCR 8250. Have attended MSA courses in General Ledger Interface, Basic Payroll, and Technical Payroll.

COMMONWEALTH EDISON
Chicago, IL

9/74 to 1/78   *Programmer/Analyst.* Official title of position was Methods Analyst. Designed, programmed, and maintained an industrial billing system based on customer demand and power usage. The system provided accounting and statistical data as well as data projections for rate changes proposed to government regulatory bodies. Programming was in COBOL and ALC with maintenance on existing systems performed in COBOL and new development performed in ALC. Approximately 25% of effort was new design. System environment was an IBM 370/158 under MVS with TSO for online access.

KRAUS VON OBERST
132 E. Chapman Avenue, #32
Tustin, CA 92682
(714) 555-2029

MARKETING ANALYST
International and Domestic

Conducted in-depth research in marketing opportunities. Developed comprehensive plans for product introduction and market penetration. Extensive knowledge of Latin American cultures and trade.

Areas of Special Emphasis

International Marketing Management
Market Research and Analysis
Product Development
Marketing Plans

B.A., Business Administration   **M.A.**, International Management

Employment Chronology

Research Analyst   1991 to Present   AmTrade International Corp.

Research foreign marketing opportunities for agricultural products in Japan. Identify Japanese tariff barriers, conduct feasibility studies, and plan for market entry. Work with contracts in commerce and government.

Teaching Assistant   1991 Academic Year   American Graduate School of
International Management

Taught course in WordPerfect, Lotus, dBASE IV, and Harvard Graphics. Wrote handbook for BASIC.

Ordering and Inventory Engineer   1981–1990   Ralph's Grocery Co.

Supervised and trained seven employees and managed inventory of over $3.5MM. Systematic inventory management, including Just-in-Time principles, resulted in inventory reduction from 37 to two pallets.

Education—Professional—Personal

EDUCATION   M.A. in International Management (concentration in International Marketing), American Graduate School of International Management, 1991

B.A., Business Administration (concentration in Marketing), California State University at Fullerton, 1990

PROFESSIONAL Member, International Marketing Association of Orange County
Foreign study at Quetzaltenango, Guatemala

PERSONAL   President, Delta Chi Fraternity. Very active in campus activities.

Sales/Product Manager (Voice Communications)

Peter S. Parker
19426 Chapaltapec Towers, #433
El Cajon, CA 92700   (619) 555-1939

EMPLOYMENT

SWITCHVIEW INC.   8/88 to Present

General Manager. Opened U.S. market for Canadian telecommunications software company. Responsible for both direct and distributor sales offices and personnel. Developed distribution agreements for Southwestern Bell Telecom, Centel and U.S. West.

FUJITSU BUSINESS COMMUNICATIONS   8/85 to 8/88

National Sales Training Manager. Created and developed training programs for all new and existing voice and data products for entire sales force, dealer network, and Bell companies. Training programs included pre-class material, leader's guides, train-the-trainer sessions, applications selling, and all tests and evaluations.

TRILLION   2/84 to 4/85

Director of Sales and Marketing. Responsible for the development and implementation of the Sales and Marketing Department. Included coordinating, developing, and implementing sales lead generation plan, opening offices nationwide, hiring, training, and managing nationwide sales force and inside staff, budget management, sales forecasting, and quota attainment.

COMDIAL   8/83 to 2/84

Product Manager—New Telephones. Responsible for the successful implementation and coordination of the new telephones offered by Comdial. Team leader and major interface between Marketing, Engineering, and Production. Company moved operations to the East Coast.

AT&T/PACIFIC TELEPHONE   1970 to 1983

Sales Manager. Responsible for teaching the sales force all aspects of voice and data communications and the System Selling curricula, including financial analysis, business problems, selling techniques, telemarketing, sales strategies, and presentation skills. Ranked top in Statewide Training Staff.

Account Executive/Communications Systems Representative. Responsible for market account plans, strategies, financial analysis, communications proposals, and sales presentations. Sold, designed, and implemented voice, data, and telemarketing systems. Responsible for revenue of over $14 million.

EDUCATION

Bachelor of Arts, California State University, Fullerton, 1970. Major in Marketing.

MILITARY

U.S. Naval Reserves, Honorable Discharge, April, 1969 to April, 1975.

REFERENCES

Provided upon request.

RICHARD M. WARDENBERG
1524 North Forest Glen Street
Torrance, CA 91799
(714) 555-9876

Objective	A challenging position with a progressive company where my analytical and technical skills can be utilized.
Experience	Arrowhead Drinking Water Company • Monterey Park, Ca.
07/87 to Present	*Senior Systems Analyst*
12/86 to 7/87	*Programmer/Analyst* Primary responsibility was the analysis, design, programming, and implementation of an online/batch route maintenance and accounting system. Additional responsibility was the maintenance of ongoing production systems.
01/85 to 12/86	Moen Computer Services Inc. • Long Beach, Ca.
	Programmer/Analyst Job responsibilities included conversion of an Inventory Control and Route Management system from a mini computer to a mainframe and design of Remittance Processing software for check processing centers.
09/74 to 01/85	Builders Emporium • Irvine, Ca.
	Assistant Manager Responsibilities included Inventory Control, Personnel Management, Customer Relations, and Employee Training.
Education	B.S., June 1980, Biological Sciences University of California, Irvine
Hardware and Software	IBM 4341, 3083 operating under MVS, DEC 11/23 CICS command level, Cobol, TSO/ISPF, DYL 280
Affiliations and Interests	Volunteer Scouter, Boy Scouts of America, Woodworking, Skiing
References	Available upon request.

Terry G. Malley
1344 Avenida de Rio • Signal Hill, CA 90888
(310) 555-5648

PROFILE

Ability to coordinate through established team effort toward the pursuit of customer satisfaction. Evaluation techniques and administrative controls motivate and perpetuate a consistency in finalization and its expected results. In being highly adaptive and creative, I use my varied experiences to properly plan and attain sought goals.

EMPLOYMENT HISTORY

1989–1990: Hell Graphic Systems • Compton, Ca.
Area Service Supervisor/Service Administrator Responsible to "positively" motivate 12 field technicians and increase customer maintenance contract base.

1988–1989: NEC America, Inc. • Hawthorne, Ca.
Pager Repair Supervisor Directed all repair activity in managing a staff of 30, as well as instituting evaluation of their individual performances on a monthly basis.

1987–1988: Eaton Corp. • Culver City, Ca.
Site Supervisor Directed 11 field engineers in a third party service environment for the repair of Digital Systems.

1982–1987: Hamilton/Avnet • Culver City, Ca.
Production Manager (1984-1987) Created a staff and governing procedures of a start-up Value Added Center for the production of custom made computer products.
Dept Supervisor/Service Coordinator Prepared and monitored new order placement and their off warranty repair.

1973-1982: Anderson Jacobson, Inc., New York, N.Y.
District Service Manager (1979-1982) After four promotions from an Associate Service Technician, I ultimately directed all repairs and refurbishment activity. In directing 14 employees, we were responsible to a customer base of 2400 terminals and 10,000 coupler/modems in New York City.

EDUCATION

Albert Merrill School, 1973; Electro-Mechanical repair of IBM equipment.
Brooklyn College, 1971-1973; Mathematics major/Psychology minor.

MAJOR ACCOMPLISHMENTS

- Increased overall production output by 32%, in managing a national depot repair facility, supervising 30 employees.
- Implementation of monthly interviews in a depot production environment, created an atmosphere where attrition was less than 8% in a 21 month span. In that same timeframe production grew at a rate of 600% (400K to 2,500K a month).
- As supervisor in the 2nd largest service center with 13% of the national terminal population, I was responsible for drastically reducing our average call downtime from 7.4 to 3.7 hours.
- In my tenure as a District Service Manager for peripheral repair, I reduced our attrition rate by 90%, overtime by 85%, and absenteeism by 50%, after my promotion from within the ranks.
- Developed, implemented, and assisted in the preparation of (OEM) product and governing procedures to end users and resellers in excess of 250K per week in sales.

References furnished upon request.

Steven M. Crawford
38488 Seaview Terrace
San Clemente, CA 92670
(714) 555-1357

OFFERING

Prime skills in all levels of management, sales, motivation and development of personnel. Knowledge of Management Information Systems, computers, word processing, and graphic arts industries with emphasis in desktop publishing.

EXPERIENCE

1985 to Present

National Accounts Manager • *Studio Software, Irvine, CA.* Introduced a new desktop publishing product as the company progressed from development stage to product release. Established outside dealer network nationwide and direct sales operation in southern California. Successfully implemented national Accounts Program with key companies throughout North America. Have developed marketing, pricing, and distribution strategy for upcoming products. Developed sales training and marketing strategy of direct sales force. Product began shipping in June 1985.

1981 to 1985

President and Founder • *Software Marketing Systems, San Mateo, CA.* Founded computer software packaging and documentation firm, specializing in the marketing of software through a new marketing concept. Developed client base in the Silicon Valley and San Francisco Bay area, including customers such as Micropro, LSI Logic, and Intel.

1977 to 1981

Western Regional Manager • *Stewart Systems Corporation, San Mateo, CA.* Established and managed the western region for a management information systems company. This included site location, real estate negotiation and leasing, furnishing, and staffing with sales and technical personnel. I developed an aggressive and professional marketing campaign to launch the product in the western region, and sales have increased at 40% per year. Office has expanded twice; three district sales managers and offices are now operating successfully.

1976 to 1977

Owner/General Manager • *The Mast, Oakland, CA.* Purchased seafood restaurant and reorganized entire operation. Recruited staff of 25, successfully negotiated a new independent union contract, developed supply sources, and instituted effective business procedures. Increased sales substantially while reducing costs. Established a prime credit rating and good cash flow.

1972 to 1976   ***Western Regional Manager*** • *ATF/Davidson Corp., Oakland, CA,* manufacturers of offset printing presses. Held overall responsibility of sales, service, development, and operation of dealer network. Revitalized region to produce substantial increase in size and volume. Increased market penetration, opening 12 major new installations. Gained familiarity with economic conditions in 13 western states and 3 Canadian provinces, personally visiting all areas regularly. Trained distributors in sales procedures and sound financial business methods. Developed and presented hard-hitting sales seminars. Set up technical support for region. Made forecasts for each product line and set quotas for each individual dealer. At ATF/Davidson's home office, played an active role in making decisions affecting marketing, developed national marketing procedures, and implemented them as policy. Implemented incentive programs, sales contests, cost control and reduction plans, and a program to enhance communication between Marketing and Manufacturing staffs. Took an active part in setting policy for improvements in manufacturing, new product development, distribution, pricing, advertising, and promotion.

1971 to 1972   ***Sr. Sales Representative*** • *Itel Corp., San Francisco, CA,* Information Products Division. Sold systems equipment: word processing, computer terminals, and input for data processing and phototypesetters.

1965 to 1971   ***Sr. Sales Representative*** • *Addressograph Multigraph Corp., Oakland, CA,* Sold computerized typesetting equipment. Developed seminars and shows. Served two years each as Government Representative and Educational Representative. Trained new sales representatives. Served as Branch Manager during his absences.

EDUCATION   B.A. in Management, St. Mary's College, Moraga, CA.
Dale Carnegie Human Relations Course and Sales Course; Assisted in teaching professional sales techniques with Dale Carnegie.

Completed numerous sales, business, and leadership courses.

Statistician/Quality Assurance (Military Transition)

Edward C. Rathbun

688 North Saufley Avenue   (612) 555-9300
Stillwater, Oklahoma 56000   Secret Clearance

OBJECTIVE

Training Specialist • Computer Programmer • Statistician • Quality Assurance

EMPLOYMENT HISTORY

US Army • *Nuclear Biological, Chemical Specialist*   Feb 84 to Feb 88

Using test beds, studied, tested and evaluated new and modified nuclear, biological, chemical (NBC) equipment and material upon arrival, such as: XM17 Sanitary Decontaminating Apparatus, Duel Purpose Chemical Staging Systems, and Modified M17 Series Chemical-Biological field protective mask.

Planned, directed, coordinated, and evaluated the daily activities of 20 personnel. Technical responsibilities included teaching, integrating and coordinating field operations for the 552nd Military Police Company.

Operated, maintained, and instructed classes on the following NBC equipment and material: IM 93 series and IM 174 series radiac meters; 1578 series radiac chargers; A/PRD-27 radiac meters; M43A1 Chemical Detection Unit and M42 Chemical alarm unit (M8A1 Chemical Alarm System); M256 Chemical Detection Kit; M1258A1 personal decontamination kit and M58 training personal decontamination kit; XM17 Sanator decontamination apparatus; M17 series Biological-Chemical modified mask; Decontamination solution number 2 (DS2); Supertropical Bleach (StB).

Served as primary instructor for the Fort Lewis ROTC programs. Developed lesson plans, conducted and instructed classes for ROTC and JROTC programs.

University Bookstore, Athens, Georgia   Sep 81 to Jul 83

University Printing, Athens, Georgia   Sep 78 to Jul 81

Worked in book and miscellaneous merchandise accountability.

Worked in production of pamphlet dummy layouts, operating manual printing machines; operating electronic computer photo prints; and type setting.

Trained and monitored the daily activities of assigned personnel.

EDUCATION

BS, Statistics, University of Georgia, Athens, GA 1983

Major Course of Study:
  Calculus, Differential Equations, Linear Algebra
  Statistics:
    Elementary and Theory
    Statistical Methods and Sample Survey Methods
    Statistical Analysis Systems (SAS) Programming

TERRI LYNN MARKSON
9753 S. Sandpointe
Santa Ana, CA 92733
(714) 555-2143

EDUCATION   Mount San Antonio College • Walnut, CA
1973–1976   Major in Computer Science (A.S. Degree Equivalent)
1971–1973   Major in Vocal Music

WORK EXPERIENCE

7/87-8/88   Canadian Insurance Company of California • Costa Mesa, CA

Systems Programmer • Management Information Systems

Responsible for installing and maintaining all operating systems (MVS/VM/VSE) and related software. Performance tuning/DASD balancing of a multiple VSE and CICS environment. Provide ongoing help and education to company staff. Installed a complete MVS/XA operating system and went through intensive training in order to convert from VSE to MVS. Attended IBM education seminars on MVS subjects, i.e., MVS Internals, SMP/E, CICS Basic Tailoring, etc. Excellent problem-solving skills.

Hardware: IBM 4381, 4341.

Software: VM/SP, VSE/SP, MVS/XA, JES2, TSO, CICS, SYNCSORT, PIE, VSAM, VTAM, SDSF, SMP/E, EPIC/VSE, PANVALET, PANEXEC, VMLIB, ISPF/PDF, PALLM, VSE/EXTEND.

11/76–7/87   Data Line Service Company • Covina, CA

1982–1987   *Senior Systems Programmer* • Technical Services Department

Responsible for installing and maintaining several VM systems and all VM-related products. Performance analysis/tuning of a multiple-CPU environment with over 200 users. Research/development of new products and techniques for in-house support. Provided ongoing help and education to company staff.

1978–1982   *Loan Application Programmer* • Off-line Programming Department
Developed and maintained off-line assembler loan programs for the servicing of savings and loan associations. Handled all conversion programming activities for new loan customers. Developed CMS programs and utilities to aid programmers. Provided help and education to company staff on the use of VM.

1976–1978   *Computer Operator* • Computer Operations Department

Developed operations standards for batch processing, JCL, etc. Assisted department manager in problem solving, scheduling, and interfacing with customers. Organized and coordinated work responsibilities for the distribution group.

Hardware: 3090, 4381 (MP), 4341, IBM 370 Series.

Software: BAL (370 Assembler), DCF (Script). DIRMAINT, DOS/VSE (POWER), GDDM, ISPF, INFO/SYS-MGMT, MVS (JES), PROFS, PSF (Advanced Function Printing), RSCS, SMART, VM/SP HPO, VM/XA, VSAM, VTAM, 3800 Printing Subsystem.

INTERESTS/HOBBIES

Music (singing, piano), skiing, hiking, camping, backpacking, racquetball.

References furnished upon request.

DONALD N. GLOSSMANN

1543 McIneree Trail • Huntington Harbor, CA 92600   (714) 555-9147

LANGUAGES   Assembler, PL/1, COBOL, SAS, FORTRAN, Lotus 1-2-3

SYSTEMS   MVS, CICS, ACP/TPF, GDDM, TSO, ISPF, PC-DOS, GENER/OL

EXPERIENCE

1981 to Present   A LARGE PROPERTY/CASUALTY INSURER   (11/81–Present)

Contribute to various enhancements of the Insurance system, some with significant revenue impact.

Develop an online Executive Information System using color graphics to present key operating results in a suitable form for executive decision support. Worked closely with executives and technical staff to accomplish the objectives of this project.

Initiate the use of PCs in the organization. Evaluate products, negotiate with vendors, install equipment, identify and train users, develop standards and policies. Emphasis on smooth integration of PCs into user departments and on immediacy of payback. In at least one department, actual staff reduction was made possible.

Participate in creation of a new system for annual budget creation. Program most of the system's reports using SAS.

While attached to DP training department, led introductory "DP Fundamentals" classes for end users. Rewrote a system to capture and report statistics on usage of CICS applications, resulting in sharply reduced maintenance requirements.

1981   SELF-EMPLOYED CONSULTANT/CONTRACTOR   (2/81–10/81)

Various assignments. Coordinated the planning and logistics of a very large software sale by a client company. Also member of project team for development of an automated Air Cargo rating system.

1979 to 1981   AM JACQUARD SYSTEMS   (12/79–1/81)

Shared responsibility for development of an Electronic Mail product. Received a bonus for this work.

Supervised maintenance and enhancement of an advanced word processor which was the mainstay of the company's product line. This product was rated #1 by DATAPRO user surveys for two years running, ahead of products by Wang, Xerox, Lanier, etc.

Taught programming class (assembler). Contributed to specifications for a new computer.

1979   COMPUTER COMMUNICATIONS, INC.   (6/79–11/79)

Member of project team to develop a stand-alone fax message switch on a microcoded 3705-compatible front end processor. Developed some sysgen tools and programmed some custom modifications to the operating system to support our application.

1978 to 1979   QUOTRON SYSTEMS, INC.   (3/78–5/79)

Launched the requirements definition study for an online manufacturing control system. Contributed to selection and justification of new hardware, a DBMS, and TP monitor software.

Maintained various business applications supporting Purchasing, Communications Engineering, Physical Inventory, and a specialized typesetting application.

1973 to 1978   CONTINENTAL AIR LINES, INC.   (6/73–2/78)

Ran an Information Center. Installed and supported a user-oriented report generator and data manipulation package. Taught user training classes, developed standards, and provided technical support.

Conducted a study to select a data dictionary product.

Provided technical and analytical support for an Information Systems Planning study.

Member of project team for development of a major new application in a very large, high performance online system. Designed and implemented a complex disk space management subsystem, including utilities, diagnostics, and offline support software.

Maintained an online Fare Quote/Ticketing application. Wrote numerous enhancements and extensions including a real time data base update utility which replaced a cumbersome and unreliable batch process.

EDUCATION   Loyola University of Los Angeles • 1969–1972

PERSONAL   Program Director, DPMA Orange Coast Chapter.

Technical Analyst (Business Investments)

WILLIAM S. BATSON

432 Peckamire Road

Santa Clara, CA 95200

Home: (408) 555-3951

Office: (408) 555-3999

SUMMARY

Creative investment analyst with particular skills in computer analysis and administration. Proven ability as a problem solver in development of computer applications and techniques useful throughout a business setting. Adaptable and self-motivated thinker who can work perceptively and constructively in a variety of situations. Broad-minded learner, particularly when working with technically-oriented subjects. Quick grasp of new subjects and an eager learner of new skills.

Key Accomplishments

- Developed computer models that allowed the quick assessment of potential investments and the impact of different deal structures.
- Administered computer system for entire company (20 people), including system purchase, maintenance, and training.
- Created extensive database application used for academic scheduling, generating over four hundred schedules.
- Took charge of Corporation's Western office, including attending monthly Board of Directors meetings and setting up of new office.

EXPERIENCE

1985–1991

INVESTORS CENTRAL MANAGEMENT CORPORATION

Assistant Vice President Asset manager for $70 million of joint-venture investments of ICM Property Investors Incorporated (ICMPI), a NYSE-listed real estate investment trust. Responsibilities included designing and setting up of asset management systems and ongoing investment management, including monthly written and oral reports to Board of Directors. Used computers extensively for analysis and office administration. Investment analyst for proposed equity investments of ICMPI and other clients. Since 1990, in charge of the Corporation's Western Office.

1984–1985

Computer Consultant Projects included academic administrative software programming and systems design and training, all on IBM personal computers.

1983

TSENG MINING COMPANY

Mill Superintendent for a gold and silver extraction project. In charge of about 20 employees.

EDUCATION

Middlebury College—B.A. in geology, minor concentrations in mathematics and Asian studies.

The Hotchkiss School graduate. Advanced work in physics and mathematics.

OTHER SKILLS

Skilled in a variety of personal computer applications, including Lotus 1-2-3, WordPerfect, Excel, Word for Windows, and various vertical market real estate applications. Extensive macro programming in a number of these applications. Familiar with a broad range of other computer applications.

BRADLEY B. DONOVAN
7223 East Valley Way • San Raphael, California 95222 • (408) 555-2030

OBJECTIVE

A technical product or product support management position in a high technology company.

SUMMARY of QUALIFICATIONS

❑ Fourteen years experience with Amdahl and IBM Corporations in product development and product support working on a diverse range of mainframe computer projects.

❑ Extensive background as key participant in PCM product development, project management, product certification and field introduction of mainframe processors involving international operations and travel. Activities include a strong exposure in 370 architecture mainframe development, testing, shipping and support. Organize worldwide product support.

PROFESSIONAL EXPERIENCE

Amdahl Corporation • Sunnyvale, CA   1980–Present

TECHNICAL PRODUCT MANAGER (1988–current)

Manage corporate international large budget projects. Involved in development of 370 architecture mainframe product involving negotiating and agreements for technical requirements by coalitions of industry leaders representing diverse groups including manufacturing, marketing, engineering, customer service, and technical publications. Also work with partner Fujitsu Corporation negotiating contracts and engineering activities. Responsible for $7.2M budget.

Represented Corporation in contract negotiations with Fujitsu focusing on engineering, quality, and technical issues. Received commendations for being an excellent negotiator. Also fulfilled role of temporary Business Product Manager and backup for a major project providing product planning, budgeting, and program momentum.

PROJECT MANAGER, PRODUCT CERTIFICATION (1986–88)

Created and managed programs to obtain design assurance and documentation for products out of development and through all phases of testing to First Customer Ship to ensure excellent RAS. Led groups and oversaw budget of $3.5M developing support and certification for all aspects of master planned projects encompassing variety of manufacturing, engineering, product support, and marketing issues. Developed effective certifications for projects including the 5890 and 5990 processors. Strategized with senior management to develop broad range of options concerning critical development, test, and field support issues impacting corporation.

FIELD DATA ANALYSIS MANAGER—HEADQUARTER SUPPORT CENTER (1985–86)

Administered site, operations, and machine audits when machine reliability and product support adherences fell below acceptable levels. Managed auditing teams in the business, government and international sectors including accounts such as First Boston Corp. Hughes Aircraft, DST Corp., South Central Bell, Social Security D.C., ATT Piscataway, Legal & General London, SECV Australia, CEGB London, and Hong Kong Shanghai Bank.

Developed and managed field monitoring programs to identify, evaluate, and resolve key problematic issues. Designed quality assurance programs based on measuring and tracking failure prone areas including design, field, and manufacturing practices. Designed and managed successful program to identify and address logistic problems with parts supply from reductions in availability to long term supply stability. Orchestrated effective project to reduce DOA and NTF rates significantly. A savings of $3.2M was realized within the first six months of operation. Reported weekly status to all applicable VPs, the COO and CEO.

TACTICAL SENIOR STAFF ENGINEER/AMDAC (1983–85)

Responsible for worldwide travel to troubleshoot accounts which were experiencing unsatisfactory MTBF levels where local and remote support were unable to resolve the problem. Functioned as AmDAC specialist to be a primary contributor on the early 580 program and served on initial install teams for each major Machine Level release.

Managed a group of nine engineers to develop a system for capturing and analyzing system failure data. Developed the 580 FASTPATH program designed to provide Field Engineer Specialists and Senior Engineers with specialized product training. Over course of this program managed ninety-five engineers.

STAFF FIELD SPECIALIST/ADVANCED SYSTEMS DEVELOPMENT (1982–83)

Performed duties of product support liaison to Engineering for hardware diagnostic packages. Developed support tools and maintenance procedures. Worked closely with customer service area and was transferred from Chicago as a Field Specialist.

FIELD ENGINEER SPECIALIST (1980–82)

Maintained accounts for 470 products in the Chicago area involving account management and product support.

International Business Machines   1977–1980

SENIOR CUSTOMER ENGINEER (1977–80)

Maintained accounts specializing in account management and product support. Other responsibilities included regional diagnostic coordinator.

EDUCATION/TRAINING

Organizational Management Development (Project Management)
ICS "Doing Business with the Japanese"
ICS "Negotiating with the Japanese"
Management Development Training
Interpersonal Communication (Human Element Seminars)

MARY ELLEN THOMPSON

9455 Barlettsville Rd. • Seal Beach, CA 92677
Home: (714) 555-9384 • Office: (714) 555-2847

OBJECTIVE

A marketing-oriented position in product and technical training.

4/87 to Present

Ashton-Tate • Torrance, CA

Training Specialist, Office Automation

Responsible for conducting training classes in MultiMate Advantage II, Framework II, Banyan Network User/Administrator and for providing technical support and software installations for in-house users. Wrote courseware for Banyan Network classes.

4/86 to 3/87

Self-employed, Yacht Chartering Service • British Virgin Islands

2/84 to 4/86

MultiMate International • East Hartford, Connecticut

Associate Product Manager, Product Development

Responsible for planning, designing, and implementing new products. Developed promotional programs, designed documentation and collateral pieces, and acted as tradeshow representative.

Technical Support Representative

Responsible for end user support in troubleshooting software and hardware malfunctions, writing procedures and program documentation, training technical support staff, and assisting with new product testing.

11/82 to 2/84

Advest, Inc • Hartford, Connecticut

Accounts Manager

Responsible for reviewing clients' account portfolios and coordinating research for problem resolution between brokers and investors.

TECHNICAL KNOWLEDGE

Includes IBM (PC, XT & AT), Macintosh, PC/MS DOS, Norton Utilities, Banyan Network, MultiMate Advantage II, Framework II, dBASE III PLUS, Byline, and most printers.

EDUCATION

Pursuing undergraduate degree in Computer Science with Training emphasis. Supplemental Education
American Management Association, "Train the Trainer"
Ashton-Tate; "Managing for Excellence"

References available upon request

SHERIL C. BOSSMANN
2279 West Tercero
Torrance, CA 90400
(310) 555-4378

OBJECTIVE A Senior Management position in the field of Financial Systems or General Management with a progressive organization.

EDUCATION MBA • Pepperdine University, 1975
Concentration: Management

BS • California State University, Los Angeles, 1972
Concentration: Business Administration and Information Systems

EXPERIENCE

7/85 to Present NORTHROP ADVANCED SYSTEMS DIVISION • Pico Riviera, CA

Training Specialist Responsible for the planning and development of company-sponsored training courses within those discipline areas established at the time of hire. As assigned, may assist and/or direct the activities of other staff personnel engaged in development, implementation, and conduct of new or ongoing training programs. May be assigned to conduct, analyze, and prepare reports and/or forecasts reflecting division training requirements. Participates in development of budget recommendations for material, equipment, and related expenses. Taught the following courses:
   • TSO/ISPF • CLIST • Beg SAS • JCL • Tellagraf

8/82 to 4/85 MAGNAVOX SYSTEMS & ELECTRONICS COMPANY • Torrance, CA

Senior Systems Analyst Performed systems analysis utilizing IBM 3033/CICS/COBOL/JCL/CMS/UFO/ADRS/FOCUS. Performed user training, walk-thru, installs, and documentation (procedures guide, user guide, system overview, hardware/software specifications) for Payroll and Personnel systems (on-line, batch). Approved Vendor Reporting System and Quality Assurance Workmanship System. Performed design of Hardware Security System and Documentation Control System for tracking top secret documents.

9/77 to Present LOS ANGELES COMMUNITY COLLEGE DISTRICT • Los Angeles

Hourly Instructor Currently, I also teach Business Administration courses at Southwest Junior College. I have taught the following subjects: Introduction to Business Data Processing, Business Systems Design and Analysis, Fortran IV, Assembler, Business Math, BASIC, and Office Machines.

11/81 to 5/82	TICOR • Los Angeles, California
	Senior Systems Analyst Was project leader for title policies, escrow, foreclosures, and related systems utilizing COBOL/WANG/DOS/IMS.
7/77 to 11/81	SECURITY PACIFIC NATIONAL BANK • Los Angeles, CA
	Assistant Vice President On a project basis, review, analyze, and evaluate new and existing systems, recommending changes to ensure management with timely, accurate, and cost efficient reporting systems. Follow up on modifications to reporting systems, ensuring necessary information is provided on a timely basis. Review Systems offered by outside vendors to determine their suitability for the application intended and evaluate alternatives.
5/76 to 7/77	LOCKHEED AIRCRAFT CORPORATION • Burbank, CA
	Sr. Financial Systems Analyst Investigated and analyzed for new and improved financial systems and assisted in preparing proper material for presentation to management. Worked with the various affected finance and data processing personnel to ensure consideration of all related system interfaces.
	Assisted in developing the design and testing of approved changes to Finance Systems. Prepared written plans, flow charts, and reports depicting detail requirements and interfaces with other finance and company systems. Provided the necessary liaison between finance and the various data processing organizations during the implementation and follow-up phases.
11/74 to 2/76	SOUTHERN CALIFORNIA GAS COMPANY • Los Angeles, CA
	Auditor Reviewed the adequacy of controls for the protecting of the company's assets and vital records. Determined that company records were prepared and utilized as specified in company procedures. Have had special assignments in tax accounting and with outside independent auditors.
8/73 to 11/74	SOUTHERN CALIFORNIA GAS COMPANY • Los Angeles, CA
	Accountant Analyzed and billed company projects from a cost accounting standpoint.
3/72 to 8/73	SAN DIEGO GAS AND ELECTRIC • San Diego, CA
	Associate Accountant Extensive management accounting training in Payroll, Accounts Payable, General Records and Report Sections.
REFERENCES	References upon request.

PAULINE SMOAK

766 W. Vallejo   (714) 555-4321 (home)
Fullerton, CA 92555   (714) 555-8655 (work)

PROFESSIONAL OBJECTIVE

A position as a Technical Writer which utilizes my communication skills, research experience, and computer knowledge.

EMPLOYMENT HISTORY

Sept. 1984 to Present

CIE TERMINALS, INC.
2505 McCabe Way
Irvine, CA. 92714

Technical Writer

Research and prepare user manuals for video display terminals, graphics terminals, and a laser printing system. Includes gathering technical information, preparing written texts, and coordinating layout and manual organization.

August 1979 to September 1984

ATV SYSTEMS, INC.
2921 So. Daimler
Santa Ana, CA. 92711
(714) 261-2390

Technical Writer   Apr. 1984 to Sept. 1984

Research and develop documentation for computerized restaurant systems and equipment. Wrote software user manuals for a chicken restaurant chain and fast-food restaurants in Europe.

Computer Librarian/Systems Coordinator   Dec. 1981 to Mar. 1984

On a Jacquard computer system, classified and cataloged data from disks for future use. Maintained files of current and previous versions of programs, listings, and test data. Prepared programs for formal release.

Tested DOS software (Documentor 3200) and hospitality systems software (FOX II Back Office System) for end user release.

Administrative Assistant   Aug. 1979 to Nov. 1981

As an assistant to the vice-president of the Systems and Software Department, I supervised and trained secretaries on word processing techniques and equipment: input specifications, technical manuals, reports, and correspondence on word processors. Interfaced between field staff and programming staff.

Sept. 1975 to Aug. 1979   AMERICAN EXPRESS COMPANY
                                             888 N. Main Street, Suite 804
                                             Santa Ana, CA. 92701
                                             (714) 547-7526

Senior Field Customer Service Representative
Interacted with Cardmembers and the establishments that accepted the
American Express Card. Maintained records of all A/E establishments
for three counties and handled secretarial duties for the Account Executive and the Territory Manager. Other duties included processing
agreements and maintaining office inventory.

HARDWARE/SOFTWARE EXPOSURE

Cie Terminals: CIT-500, CIT-101eg, CIT-420, CIT-220+ and CIT224
Jacquard computers: J100, J300, J500 and J425
Documentor POS Systems, FOX II Back Office Systems
Apple Macintosh: MacDraw, MacPaint
Basic programming, Type-Rite Word Processing, WordStar, and Word 11
Word Processing

EDUCATION

A.A.—Liberal Arts, Saddleback College • Irvine, CA.
Graduated with honors. Courses included Chemistry, Biology, Psychology,
and Basic programming.
Currently attending California State University • Fullerton.

ARTHUR THOMAS STEWART
23111 Paxatawnee Way
Aliso Viejo, CA 92699
(714) 555-2828

GOAL	Telecommunications: Technical and Administration.
GENERAL SUMMARY	Responsible for telecommunication system analysis, technical systems design, equipment, line service, voice and data transmission. This includes planning, budgets, long distance cost control, software translations, troubleshooting, customer service and automated system administration. Effective in reducing annual cost of system operation each of the last five years while upgrading and doubling system size.

BUSINESS EXPERIENCE

1985 to 1991	UC-Irvine Medical Center • Orange, CA.
	Telecommunications Manager. Responsible for systems analysis, design, order activity, planning, budgets, long distance cost control, problem-solving and supervision of personnel. Effective in reducing annual cost of operation each of the last five years while upgrading and doubling system size. Added Voice Mail and Computerized Operator Consoles reducing the operator load by 40%. Handled various special projects including methods and procedures, planning, personnel management and implementation.
1983 to 1985	AT&T Information Systems • Santa Ana, CA.
	Technical Sales/Service. Responsible for System 85 implementation for major accounts. This included design, software translations, and equipment. Handled first cut-over of System 85 without a single problem. Additional responsibility for marketing administrators, sales, service, and order entry personnel handling embedded telephone equipment.
1958 to 1983	Pacific Telephone Company • Orange, CA.
	Involved in all aspects of telecommunications including voice and data, marketing, line service, sales, account service, design, analysis, equipment management, training, and lecture demonstrations. Worked with both large and small business.
SPECIAL TRAINING	AT&T and Pacific Telephone: Multiple technical and management seminars and training courses. Radioman Class "A," Submarine.
PERSONAL	Enjoy reading, science, politics, and woodworking.
REFERENCES	Available upon request.

ANDREW F. BARKOWICZ
(714) 555-6767

2438 Amarillo Avenue
La Palma, CA 90777

TELECOMMUNICATIONS MANAGEMENT
EDP/MIS
Office Automation . . . Data Communications
Project Management . . . Product Development
Customer Service Manager

QUALIFICATIONS

Twenty-two years of versatile telecommunications management experience with Pacific Telesis and Hughes Aircraft Corporation in a variety of assignments to include direct customer support services, building positive customer satisfaction through design of new trouble-shooting procedures and equipment, complemented by BS in Computer Science, Cum Laude, National University, Irvine. Key strengths include:

- Troubleshooter—on equipment and procedures to enhance operations.
- Customer Service/Relations—enhanced business by providing technical assistance, training and support.
- Training Customers/Technical Staff—instructed on equipment and procedures to maximize operations and reduce costs.
- Designed new systems—for internal and customer use.

REPRESENTATIVE ACCOMPLISHMENTS

Resourceful Problem Solver—As District Service Manager evaluated cost efficiency and performance for the district. Found little to no measurement capability for management. Recommended development of computer-assisted procedures to monitor operations and field service reports. Utilization of personnel increased from 23% to 41% without overtime and 37% to 53% with overtime. Modified procedures for shipping equipment by field service engineers resulting in reduced employee accidents and elimination of damage to equipment costing $200M. As supervisor/section head realized a 15% improvement in operations over six months by implementing automation processes to track orders and installation costs.

Designed new system and implemented a statewide Order Tracker and Report Generator System for senior management that yielded an estimated $10MM savings in operating costs for more than two years. Also implemented a project that reduced the average trouble ticket duration from 30 hours to 3.5 hours for a telecommunications network linking 22 computer centers.

Managed, designed, and implemented a "Total Information Center" (office auto-mation project for a staff of 20,000) for Switched Services of Southern California.

- Assisted in Cost justifications.
- Order placement/tracking of hardware/software.
- Set-up, inventory and initial user training.
- Follow-up extensive training in:
  — Spreadsheets
  — Word Processor
  — Graphics and Communication
  — Data Base Procedures

Also developed special IBM-PC software and prepared training documents for the user. Improved operations by 3.0% provided a reduction of 30 staff members for a savings of $1.2MM per annum.

BACKGROUND

Network Equipment Technologies   Dec. 1988 to present
   District Service Manager

Hughes Aircraft Corporation   Sep. 1986 to Dec. 1987
   Supervisor, Corporate Telecommunications

Pacific Bell   Feb. 1970 to Aug. 1986
   Specialist, Data Technical Support

Hughes Aircraft   Dec. 1968 to Jan. 1970
   Technician
   Electronic

U.S. Army   Honorable 1966–1968

References will be furnished upon establishing mutual interest.

GEORGE PHILLIP HEIMAN
(619) 555-8456

3985 Punta Playa
Vista, CA 92476

MANAGEMENT INFORMATION SYSTEMS
Strategic Planning • Telecommunications • System Analyses/Design • User Liaison

Eight years management experience in telecommunications and information systems highlighted by the management of a major merger of telecommunications and computer systems management for home office involving 1,300 people and 26 geographically separate locations involving an additional 7,500 people, Features supervision of system software development, quality control and testing procedures. Includes writing of critical analyses, policies and procedures. Complemented by an MBA in finance and accounting and the following skills and qualities:

situation analysis
distributed data processing
directs meetings skillfully
overhauling ineffective methods
works closely with top management
wins cooperation from people at every level
focuses others' energies towards solutions
source of new ideas that work
easily wins peoples' confidence
"goal-oriented" problem solving
ability to get things done
skilled versatile writer
entrepreneurial spirit

EXAMPLES OF EFFECTIVENESS

- Initiated new test procedures for operating system software for multi-million dollar communications computer system. Reduced testing time by 25%. Updated software and went on-line without causing subsequent problems.

- Developed integration plan that streamlined the management, operation and long range planning of information systems at 26 separate locations. Concept adopted company-wide. Cut manpower requirements by three people at each location and saved $300M+.

- Prepared critical analyses, changes for new policies, regulations, and operation concepts, including procedures to ensure budgeting of communication and computer systems requirements for new construction.

- Wrote speeches for senior managers addressing meetings and seminars of information systems, communication systems, and computer systems professionals.

- Built a successful insurance business from scratch. Assisted more than 125 clients to determine investment, estate and disability needs, provided appropriate financial products solutions.

RELATED BACKGROUND

NEW YORK LIFE INSURANCE CO. • Omaha, NE   1987–1990
   Agent

COLLEGE OF SAINT MARY • Omaha, NE   1986–1988
   Adjunct Instructor
   Taught Distributed Data Processing

UNITED STATES AIR FORCE   1980–1987
   Headquarters Strategic Communications Division
      Offutt Air Force Base, NE   1984–1987
         Plans and Programs Staff Officer
      Reese Air Force Base, Texas   1983–1984
         Deputy Class Leader and Student Pilot
   1st Aerospace Communications Group
      Offutt Air Force Base, NE   1981–1983
         Chief, Executive Programming Section
      Keesler Air Force Base, MS   1980–1981
         Class Leader and Student
     • Learned programming skills and technical elements of managing communications systems for U.S.A.F.

EDUCATION

M.B.A., Finance and Accounting • Creighton University, Omaha, Nebraska, 1986

B.A., Chemistry • College of St. Thomas, St. Paul, Minnesota 1978

Rudy P. Echivara
9876 Prospect • Fontana, CA 92500
(714) 555-9812

GENERAL SUMMARY

Master's in Computer Science (equivalent) IBM's Systems Research Institute. Extensive background in systems engineering, computer programming, training, test planning, test procedures, project management, and customer service. Familiar with Fortran, COBOL, Assembly Language, CMS-2, Basic, Pascal, software test, systems test and systems integration.

BUSINESS EXPERIENCE

1965 to 1989   International Business Machines (IBM)

1986   *System Test Engineer* • Boulder, CO.

Installed, maintained, and provided technical direction and support to department personnel. Responsible for up to 30 PCs used as smart terminals to access a main frame computer. Member Systems Test Group responsible for systems test and integration of ground based support systems for a classified U.S. Air Force program.

1976   *Advisory Programmer* • Westlake Village, CA.

Team leader for wide variety of commercial and governmental projects including communications interface design for *New York Times* information system; design and proposal efforts for the Ohio State University automated library circulation system; system architecture and production management of the Integrated Radio Room of the Trident Submarine; design, implementation, and integration of the information sub-system for the U.S. Navy Cuttysark-class ships.

1965   *Advisory Analyst*

Provided technical support as a member of the FAA National Account Marketing team. Chief programmer for Data Acquisition Network at U.S. Weather Bureau. Responsible for the design, curriculum, installation, implementation, and training programs.

Prior • Field and Systems Engineer. Installed and maintained IBM Unit Record equipment and large scale digital computers.

EDUCATION

Graduate   Master's in Computer Science (Equivalent). Highest Honors Systems Engineering. IBM Systems Research Institute.

IBM: Management School and various technical trade schools.

PERSONAL

Enjoy ham radio, computers, and volunteer work with Boy Scouts of America.
DaVinci Award: Top 50 division engineers.
Outstanding Contribution Award: Systems performance.
Outstanding Contribution Award: National Weather Bureau.

REFERENCES   Available upon request.

Section 3

Special Types of Resumes

This section contains two special resume types, the Background Summary and the Narrative. They are normally used by either search organizations or companies that are presenting you and your background to others.

The use of special resume types should be limited to circumstances where you can be effectively introduced to a prospective employer by an intermediary or third party.

Third party introduction is useful because a reputable spokesperson on your behalf lends credibility to your position. It is a refinement of the networking process. Use it whenever the opportunity presents itself. However, do not present yourself in the third person. Besides being grammatically awkward, you come off sounding corny and possibly unprofessional.

REALTIME ASSOCIATES

Peopleware - Fulfilling the Technical Equation

Background Summary
of
JUDY BRODERICK

Upon graduation from college, Judy performed an internship in an accounting position in Austria at the International Council on Monuments and Sites. This required her to learn the German/Austria accounting system (Summer 1980).

Upon her return to California, she worked in advertising, reporting to and supporting three account executives. This required project monitoring and financial oversight of the accounts (two years - 1981-82).

Judy then moved to Boston where she worked for a start-up architectural firm where she was the only administrative employee. It was here that she was introduced to computer business systems and worked with a Digital computer system. As the business of the company grew, Judy was the trainer for all computer activities (two years - 1983-84).

The next year she worked as a temporary employee in a variety of business environments. During this time she learned additional PC computer applications.

The temporary assignments led Judy to a full-time position at Princeton University where she worked for the next five years as the Systems Manager and Administrative Coordinator for two projects. During this period she was responsible for the development of computer based systems to track multi-million dollar grants and the application of data base management systems to track the proposals that Princeton was funding. In addition to designing and developing applications in LOTUS 1-2-3 and RBASE and PARADOX, Judy also planned and installed a Novell LAN. As the Systems Specialist, she trained other members of her organization in the use of spreadsheets, data bases, word processing, and the network.

Since moving back to California, Judy has worked as a contractor in various organizations in the Silicon Valley.

Judy is seeking a position where she can use her computer systems and project management skills.

EDUCATION: BA, German Studies, Mount Holyoke College, 1980

Graduate Courses in an MA program
at San Francisco State College in Systems Development
 in Educational Administration

Princeton University: Business Courses:
The Management of Innovation
Microcomputer and Information Technologies

Various Microcomputer Application Courses

SALARY: Salary at last permanent position (Princeton): $30,000

David J Moore

Consultant to Management • Author • Seminar Leader

David Moore is a nationally recognized authority and author on Human Resources Management Systems, with more than 20 years of experience in the management and development of information systems projects and a strong commitment to training and developing human resource potential.

In the area of Project Management, he developed a Project Management course for Security Pacific using the Viewpoint Project Management software. As the Director of Aviation Systems for the United States Marine Corps, he directed the NALCOMIS project (Naval Aviation Logistics Command Management Information System) involving design and development of 92 minicomputer facilities and supporting software for the management of aviation material and maintenance activities. He also directed the project that upgraded IBM mainframe Marine Corps computer systems on the West Coast and consolidated them into the first networking application.

He is the author of *Job Search for the Technical Professional* (John Wiley, 1991) and *Recruiting and Hiring the Computer Professional* (Van Nostrand Reinhold, 1987) as well as articles for publications such as *National Business Employment Weekly, Information Systems News, Data Management, Information Executive, Orange County Business Journal,* and *The Marine Corps Gazette.*

With his own company, Realtime Associates, and its training subsidiary, Pro/Skills Seminars, he has presented the Employment Dynamics Seminars to a variety of clients, including aerospace/defense, financial institutions, insurance, entertainment, public utilities, and manufacturing firms. He has spoken to International Computer Programs (ICP) workshops and at COMDEX.

He teaches in the graduate programs (M.B.A. and M.A.) of Webster University and the University of Phoenix and holds California Community College Instructor credentials in Computer Technology and Management/Supervision.

Dave received his A.B. degree in Economics from Indiana University and his M.S. degree from The American University in Technology Management with concentrations in Computer Systems and Management Information Systems. He has done additional graduate work at Catholic University in human resource psychology.

He is a member of the American Society for Training and Development, Toastmasters International, and Speakers USA.

Appendix

Additional Software

RESUME PREPARATION SYSTEMS

Resume preparation systems are scaled-down word processors that provide a variety of resume templates or formats that can be used in the design and creation of your resume. A common feature of these packages is their claim to creating the "winning resume." It must be emphasized that a "winning" or "best" resume is the one that gets you a face-to-face interview with the person who can hire you. It's what you put into the resume coupled with the way you display it that will "win" for you.

In addition to resume formats, resume preparation or writing systems often include a variety of built-in type fonts, sample job search correspondence and model letters, contact tracking systems, appointment calendars, interview tips, and career advice. Again, the choice is what will best meet your needs and personal preferences.

One example of a commercial software package for this purpose is *ResumeMaker with Career Planning* by Individual Software, Inc. The retail price of this software is $49.95.

CAREER GUIDANCE SYSTEMS

Career guidance systems attempt to address various aspects in the spectrum of the job search. This spectrum ranges from choosing a career from interests and experience to resume preparation and contact tracking. An excellent example of this type of software is *Career Design* by Career Design Software of Atlanta, Georgia. It views itself as a "complete career-planning tool." On the career planning side, it contains self-help exercises that help you decide on a career path, communicate your strengths and abilities, and guide you in your current career. It contains over 50 modules, including "Establishing Your Career Direction," "Identifying Viable Career Options," "Discovering Your Valuable Hidden Talents and Skills," "Determining Your Optimum Working Environment," "Writing Letters That Command Attention," and "Choosing the Best Resume Format" (seven are included). Also, included are interview tips, salary negotiation advice, and how to choose the best offer.

Career Design uses the job search methods of John C. Crystal, who was the major contributor to Richard N. Bolles' classic best seller on career advice, *What Color Is Your Parachute?* The retail price of this software is $99.95.

JOB SEARCH SOFTWARE AVAILABLE THROUGH SHAREWARE

An often overlooked source for high-quality resume and career software is the "shareware" marketplace. This is the "economy" route to software acquisition, because you only pay for the cost of the diskettes, ranging from $2 to $5. The concept behind shareware is "try before you buy." If you like the software, you are encouraged to send in the license fee, which usually ranges from $20 to $60. This buys you the right to use the software, a full set of documentation, technical support, and update information. It is much less than you would pay for commercially produced software, and its quality is often as good as or better than the commercial counterparts.

The following are some shareware packages available from Public Brand Software at $5 per disk. They are distributed through a catalog. Public Brand Software, located in Indianapolis, Indiana, may be contacted at (800) 426-3475.

Job Track. This is a simple data base tracking package. It stores contact information and searches by company name and date contacted. Licensing is $22.

Resume Master. This program creates and prints chronological or targeted resumes. It is menu-driven for ease of use and comes with on-line help. Licensing is $20.

Resume Professional. This program helps you organize your resume information by keeping your personal, educational, work, and reference data. It enables you to cut and paste the information to create custom and targeted resumes. Licensing is $20.

These three programs are included on a single disk for a cost of $5.

Shareware also offers excellent job search organization and career guidance software. Here are two examples:

Looking for Work. This is an interactive guide to marketing yourself for a new job. It assists in developing a job search plan, provides a timetable, and recommends specific activities. It also reviews and analyzes your financial and time resources. It provides step-by-step guidelines and criteria for developing a list of target companies and then guides you in preparing a detailed employment history and industry-specific resumes and cover letters. It also provides reminders for a week-by-week list of tasks, activities, and milestones. The cost of a single disk for the system is $5. Licensing is $25.

Job Search. This is a full-blown job search system complete with a database for tracking companies either by industry or geographical preference and a resume preparation system. The word processor comes complete with a spell checker and guides for resume, cover letter, and follow-up letter preparation. It also comes with a time manager and planner for interview scheduling and activities planning, as well as a pop-up phone dialer, calculator, calendar, and area code list. The disk is $5, and licensing is $60, which includes a personal marketing plan.

OFFICE ACCELERATOR

Office Accelerator is a Windows-based product that permits the job seeker to use a pull down menu from within his or her favorite word processor. *Office Accelerator* is a relational database that is installed as a set of commands on the menu bar of Windows software programs such as *Microsoft Word for Windows, Word Perfect for Windows,* and *Ami Pro*. From the main menu, the job seeker is able to generate letters, envelopes, faxes, labels, forms, mail merges, and phone book reports. Information on prospective employers and network contacts is maintained in the contact database, called the "Phone Book." The job seeker can create customized forms, letters, and faxes. The product also permits direct import of *dBASE* (*.dbf) files so other databases can be used. The value of *Office Accelerator* for the job hunter is the ease with which letters, phone calls, faxes, and so forth can be generated from within a Windows-based word processor without the need to go into often complex mail merge features of a data base management system.

Information regarding *Office Accelerator* is available from Baseline Data Systems Corporation, 3625 Del Amo Boulevard, Suite 245, Torrance, California 90503, (310) 214-8528.

Index to Resume Samples by Job Title

If you're not looking here, you're hardly looking.

There are lots of publications you can turn to when you're looking for a job. But in today's tough job market, you need the National Business Employment Weekly. It not only lists hundreds of high-paying jobs available now at major corporations all across the country, it also gives you valuable strategies and advice to help you land one of those jobs. NBEW is a Wall Street Journal publication. It's the leading national job-search and career guidance publication and has been for over ten years. Pick it up at your newsstand today. Or get the next 12 issues delivered first class for just $52 by calling toll-free...

800-367-9600

National Business Employment Weekly

If you're not looking here, you're hardly looking.